Health for All?

Edited by
The World Life Sciences Forum
BioVision

Read More About the Topic

William H. Foege et al. (Eds.)

Global Health Leadership and Management

2005
ISBN 0-7879-7153-7

John Bryant

Introduction to Bioethics

2005
ISBN 0-470-02198-5

Stefan H. E. Kaufmann (Ed.)

Novel Vaccination Strategies

2004
ISBN 3-527-30523-8

Guido Grandi (Ed.)

Genomics, Proteomics and Vaccines

2004
ISBN 0-470-85616-5

C. Everett Koop, Clarence E. Pearson, M. Roy Schwarz (Eds.)

Critical Issues in Global Health

2002
ISBN 0-7879-6377-1

Health for All?

Analyses and Recommendations

Edited by
The World Life Sciences Forum
BioVision

WILEY-VCH

WILEY-VCH Verlag GmbH & Co. KGaA

The World Life Sciences Forum
Scientific Foundation Biovision
Dr. Philippe Desmarescaux
4, rue Président Carnot
69002 Lyon
France

www.biovision.org

Library of Congress Card No.: applied for
A catalogue record for this book is available from the British Library.

Bibliographic information published by Die Deutsche Bibliothek
Die Deutsche Bibliothek lists this publication in the Deutsche Nationalbibliografie; detailed bibliographic data is available in the internet at http://dnb.ddb.de.

Printed in the Federal Republic of Germany
Printed on acid-free paper

Cover Design   SCHULZ Grafik-Design, Fußgönheim
Typesetting   Manuela Treindl, Laaber
Printing   betz-druck GmbH, Darmstadt
Binding   Litges & Dopf Buchbinderei GmbH, Heppenheim

ISBN-13: 978-3-527-31489-8
ISBN-10: 3-527-31489-X

The World Life Sciences Forum BioVision

As the analysis of recent decades clearly illustrates, science and technology are making exponential progress in all fields. This is particularly true of two areas: Information and Life Sciences, recognized as the keys to economic development and substantial changes in our lifestyle throughout the 21st century.

Health-related and food-related industries, like all sectors with an environmental component, are finding themselves increasingly influenced, and indeed transformed, by the progress made in life sciences.

Since its inception, BioVision has addressed many of the vital life sciences issues facing our world today, and has achieved much of what it initially set out to accomplish: mobilise foremost specialists along with policy makers, researchers, consumer representatives, patients' associations, NGOs and the media, with the aim of fostering open debate and exchange between Science, Society at large and Industry.

The unique concept of The World Life Sciences Forum BioVision was born in Lyon and will continue to thrive thanks to support from the Ville de Lyon, Grand Lyon, Département du Rhône, and the Région Rhône-Alpes.

Mr. Raymond Barre

Former Prime Minister,
Former Vice President of
the European Commission

Prof. Federico Mayor

Former Director-General
UNESCO
President, Fundación Cultura de Paz

Prof. François Gros

Honorary Permanent Secretary,
Académie des Sciences
de l'Institut de France

Co-Chairs of The World Life Sciences Forum BioVision

President,
International Scientific Committee

Health for All?: Analyses and Recommendations
Edited by The World Life Sciences Forum – BioVision
Copyright © 2005 Wiley-VCH Verlag GmbH & Co. KGaA, Weinheim
ISBN: 3-527-31489-X

Preface

Since its inception in 1999, The World Life Sciences Forum BioVision has played a vital role in promoting the sustainable development of Life Sciences on an international level, ensuring that they remain beneficial to humankind and the environment, and committed to ethics.

BioVision has established itself as a platform for dialogue and debate by engaging top stakeholders and policy makers from *Science, Society* and *Industry* in discussions of what science can do, what society is willing to accept, and what industry can produce, all within a sound ethical framework. BioVision 2005 casts current and future issues in life science within the three inter-dependent sectors of *Health, Agriculture* and *Environment.*

The BioVision debate seeks to:
- Explore major topics, identify disagreements, and reach consensus when possible.
- Identify existing opportunities and facilitate their implementation to benefit both developed and developing countries.
- Promote positive action to define priorities, identify precautionary principles, and create a path to future success.
- Build a sustained communication flow around key issues, ensuring that they are understood and shared by stakeholders and the society.

These new BioVision volumes could not have materialized without the essential contributions of a wide range of people. Special thanks are due to the BioVision Lyon team organizing and implementing BioVision 2005. Dominique Lecourt and Ben Prickril assisted in preparing introductory materials for these volumes. David Zavaglia, Christine Toutain, Cécile Trespeuch, Anne-Sophie Bretonnet, Sandra Zoghbi, and Laurence Clement of BioDocs (*www.biodocs.net/lyon*) prepared the reference materials. The *Syntheses and Recommendations* were prepared with the expert assistance of Pierre Anhoury, Jens Riese, Oskar Slotboom, Clare Cockcroft, and Radhika Bhattacharya. Daniel Leclercq has been extremely helpful in moving the books from concept to fruition. Frank Weinreich and his colleagues at Wiley-VCH

Health for All?: Analyses and Recommendations
Edited by The World Life Sciences Forum – BioVision
Copyright © 2005 Wiley-VCH Verlag GmbH & Co. KGaA, Weinheim
ISBN: 3-527-31489-X

have been exemplary in putting together the entire volumes, including transcription and editing of the presentations.

Special appreciation is due to the truly outstanding group of BioVision 2005 Chairs, all of whom have graciously donated their considerable skills to formulating and implementing the Forum.

Lyon, October 2005

Philippe Desmarescaux
Chairman and Founder
The World Life Sciences Forum BioVision

Contents

Health for All?: Analyses and Recommendations
Edited by The World Life Sciences Forum – BioVision
Copyright © 2005 Wiley-VCH Verlag GmbH & Co. KGaA, Weinheim
ISBN: 3-527-31489-X

Module I
Is the Investment in New Therapies Paying Off?

Health for All?: Analyses and Recommendations
Edited by The World Life Sciences Forum – BioVision
Copyright © 2005 Wiley-VCH Verlag GmbH & Co. KGaA, Weinheim
ISBN: 3-527-31489-X

Introduction

*Dominique Lecourt**

Progress in biomedical research has advanced rapidly in recent years. This has contributed to better understanding of diseases, new therapeutic options, and improvement in both length and quality of life. The contributions of genetics to the study and treatment of cancer illustrates this wonderfully. Recent research reveals genetic mechanisms, particularly in the early stages of development of cancer cells. As for traditional therapies (surgery, chemo-therapy, radiotherapy), new strategies based on molecular targets have shown high efficacy with minimal side effects.

But are such innovations economically profitable? The example of the development and use of vaccines for diseases like variola or measles makes it possible to understand the nature of the problem. Beyond the immediate calculations of cost-benefit ratios there is an excellent argument that children growing up in good health will contribute to the future economic development of their countries. If poor countries are not solvent today isn't there a means of enabling them to improve this situation for tomorrow? This promotes the idea that the 21st century will be the "century of the vaccines".

However, in order to succeed, neither research nor good intentions will suffice. Innovation must be supported by dynamic partnerships between public and private research organizations, and between academic researchers and biomedical industries. Governments and other decision makers under-stand implicitly the need for robust health systems, effective organization of markets and adequate regulatory structure to support innovation and the diffusion of progress through all levels of society.

---

* Professor at the University of Paris 7,
  General Delegate of the Biovision/Academy of Science Foundation.

Health for All?: Analyses and Recommendations
Edited by The World Life Sciences Forum – BioVision
Copyright © 2005 Wiley-VCH Verlag GmbH & Co. KGaA, Weinheim
ISBN: 3-527-31489-X

Author Biography

Zhu Chen

Vice President, Chinese Academy of Sciences

Prof. Zhu Chen obtained his master's degree at Shanghai Second Medical University in 1981 and doctor's degree at Paris VII University in 1989, and presently is member and Vice President of the Chinese Academy of Sciences, a Foreign Associate of the National Academy of Sciences of the USA, Titular Member of European Academy of Arts, Sciences and Humanities, Co-Chair of InterAcademy Panel, Director of the Chinese Human genome center at Shanghai and Director of the Shanghai Institute of Hematology. He is devoted to research on leukemia, in the field of which he is well-known for the advancement of molecular target-based therapy of human cancer after the breakthroughs in the clinical and molecular study of the treatment of acute promyelocytic leukemia with all-trans retinoic acid and arsenic trioxide. He is also now playing a leading role in human genome project of China. Zhu Chen was the first non-French winner of "Prix de l'Oise" by "La Ligue Nationale contre le Cancer" of France. In October 2002, he was awarded the "Chevalier de l'Ordre National de la Légion d'Honneur".

1
New Frontiers in Cancer Treatment

Zhu Chen

1.1
Introduction

Cancer is a most notorious disease which causes widespread mortality and morbidity on a worldwide basis. Moreover, the harm caused by cancer to society is becoming increasingly serious. When considering the history of disease mortality in developed countries it can be seen that, about 100 years ago in the USA for example, cancer was the cause of death in only 3.7% of the population. However, by 1997 it was responsible for 23% of all deaths in the USA, and was ranked number two among the leading causes of mortality. Likewise, in China – a developing country with a booming economy and a rapidly changing lifestyle – there has been a significant increase in mortality due to cancer over the past two decades (Figure 1.1). For example, in 1982 cancer was the third highest cause of death among the Chinese population, but today it is the leading cause.

In continuing the fight against cancer, it is essential to understand the major biological features of cancer cells. These cells differentiate themselves from normal cells in several ways: a significant growth advantage caused by uncontrolled cell division and the arrest of differentiation and maturation; a deregulated cell death leading to an accumulation of malignant cells; and an ability to invade surrounding tissues and to metastasize to new body sites (Figure 1.2).

Clinically, different stages of disease progression can be either observed or detected, from precancerous status to localized tumor to regional tumor and to metastasis. Since the 1970s, two types of gene have been identified as being implicated in the oncogenetic process. A large body of evidence suggests that loss of function in tumor suppressor genes and gain-of-function mutations in oncogenes categorize most, if not all, human cancers. On occasion, genetic abnormalities can occur at a juvenile level, and this gives rise to a predisposition

Health for All?: Analyses and Recommendations
Edited by The World Life Sciences Forum – BioVision
Copyright © 2005 Wiley-VCH Verlag GmbH & Co. KGaA, Weinheim
ISBN: 3-527-31489-X

A

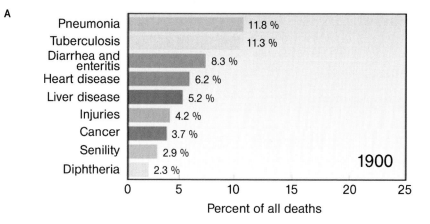

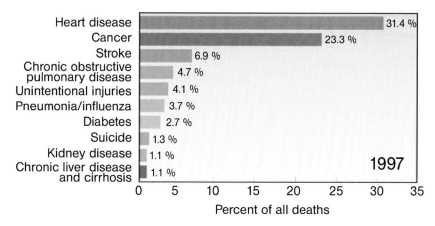

Figure 1.1 Contribution of cancer to human mortality.
(A) Leading causes of death in the USA.
Source: Centers for Disease Control and Prevention, National Center for
Health Statistics, National Vital Statistics System and unpublished data
(data from 1900 do not represent all states).

B

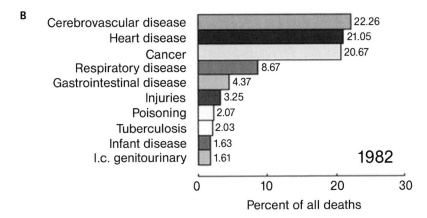

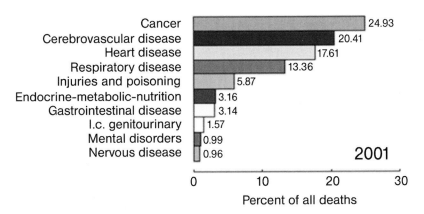

Figure 1.1 (Continued)
(B) Leading causes of death in China (data are from Chinese cities).

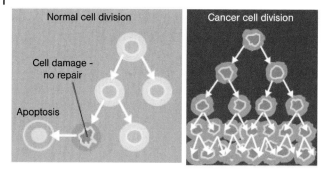

Cancer progression

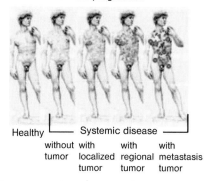

Healthy — Systemic disease —

without	with	with	with
tumor	localized	regional	metastasis
	tumor	tumor	tumor

Figure 1.2 Cancerogenesis and progression.
Malignant neoplasms are characterized by the proliferation of anaplastic cells
that tend to invade surrounding tissue and metastasize to new body sites.

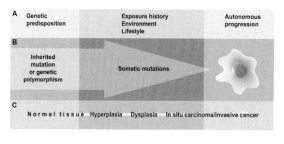

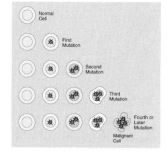

Figure 1.3 Disease pathogenesis of cancer.
(A) Factors contributing to the generation of the cancer cell.
(B) The effects of these factors on the cell.
(C) The pathologic outcome of (A) and (B), with the progression
from normal tissue to invasive carcinoma highlighted.

to cancer. Nevertheless, in most cases the cancer is associated with somatic mutations as a result of an interaction between environmental factors and genetic information (Figure 1.3). An accumulation of gene mutations has been shown to accompany disease progression underlying an increased degree of malignancy.

1.2
Cancer Treatment

Conventionally, cancer was treated either surgically, or by chemotherapy and/ or radiation therapy. These therapies have evolved over time, with surgery having now become more precise and less invasive.

Although chemotherapy was first developed during the mid-1940s, since the 1960s the combined use of drugs such as antimetabolites, alkylating agents, topoisomerase inhibitors and anticancer antibiotics has greatly improved the efficacy of chemotherapy (Table 1.1). Likewise, sophisticated instruments have been developed to enhance the effect of radiation therapy. Despite these improvements, however, the outcome of conventional cancer therapy remains far from satisfactory. Unfortunately, neither chemotherapy nor radiotherapy is able to distinguish normal cells from cancer cells, and consequently these agents will not only kill cancer cells but will also cause damage to normal cells and tissues. In clinical terms, the toxicity of these therapies often reaches the maximum tolerated by the organism.

Some cancers are insensitive to both chemotherapy or radiotherapy, however, and consequently new concepts of cancer therapy require the global decoding of genetic information. This will allow the genome-wide characterization of molecular abnormalities in cancer which, in turn, will permit the development of new therapeutic strategies. In fact, this was the initial motivation, some 19 years ago, for the Human Genome Project to be initiated.

Following completion of the Human Genome Project, and the near-completion of the SNP haplotype, a new project was recently launched to address the cancer genome. Recent advances have shown cancer to be a disease which involves dynamic changes in the genome and, indeed, it is estimated that about 600 genes are involved in the process of oncogenesis (Figure 1.4). These genetic events confer cancer cells with a self-sufficiency of growth signals, an insensitivity to antigrowth signals, a capability to evade apoptosis, a limitless replicative potential, sustained angiogenesis, and properties of tissue invasion and metastasis. Clearly, the elucidation of the genetic basis of cancer will not only shed new light on carcinogenesis, but also provide novel therapeutic perspectives (Figure 1.5).

Table 1.1 The history of chemotherapy (1942 to present).

Year	Event
1942	Louis Goodman and Alfred Gilman use nitrogen mustard to treat a patient with non-Hodgkin's lymphoma and demonstrate for the first time that chemotherapy can induce tumor regression.
1948	Sydney Farber uses antifolates to successfully induce remissions in children with acute lymphoblastic leukemia (ALL).
1955	The National Chemotherapy Program begins at the National Cancer Institute (NCI), a systematic program for drug screening commences.
1958	Roy Hertz and Min Chiu Li demonstrate that methotrexate as a single agent can cure choriocarcinoma, the first solid tumor to be cured by chemotherapy.
1959	The Food and Drug Administration (FDA) approves the alkylating agent cyclophosphamide.
1965	Combination chemotherapy (POMP regimen) is able to induce long-term remissions in children with ALL.
1970	Vincent DeVita and colleagues cure lymphomas with combination chemotherapy.
1972	Emil Frei and colleagues demonstrate that chemotherapy given after surgical removal of osteosarcoma can improve cure rates (adjuvant chemotherapy).
1975	A combination of cyclophosphamide, methotrexate and fluorouracil (CMF) was shown to be effective as adjuvant treatment for node-positive breast cancer.
1978	The FDA approves cisplatin for the treatment of ovarian cancer, a drug that would prove to have activity across a broad range of solid tumors.
1989	The NCI introduces 'disease-oriented' screening using 60 cell lines derived from different types of human tumor.
1992	The FDA approves paclitaxel (Taxol), which becomes the first 'blockbuster' oncology drug.
2001	Studies by Brian Druker lead to FDA approval of imatinib mesylate (Glivec) for chronic myelogenous leukemia, a new paradigm for targeted therapy in oncology.
2004	The FDA approves bevacizumab (Avastin), the first clinically proven anti-angiogenic agent, for the treatment of colon cancer. Researchers at Harvard University define mutations in the epidermal growth factor receptor that confer selective responsiveness to the targeted agent gefitinib, indicating that molecular testing might be able to prospectively identify subsets of patients that will respond to targeted agents.

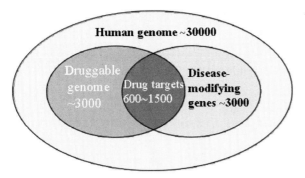

Figure 1.4 The impact of genomics.
(Source: Hopkins and Groom, Nature Rev. Drug Discov. 1, 727, 2002).

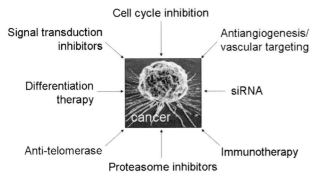

Figure 1.5 Treatment of cancer in the era of genomic medicine.

1.3
Target-based Therapies

Recently emerging anticancer strategies have been mainly based on "molecular target-based therapy". The target chosen is critical to the malignant phenotype, but is not expressed in vital organs and tissues, thereby conferring a high efficacy with minimal adverse effects. The first examples of targeted therapy were provided by breakthroughs in the hematological setting (Figure 1.6). For example, leukemia is not a single disease but rather a group of diseases including myeloid and lymphoid leukemia subtypes; these can be further divided into acute and chronic subtypes. Importantly, specific gene mutations such as chromosome translocations and point mutations have been identified as playing a key role in leukemogenesis, thus providing potential targets for specific treatment to be developed. The first opportunity to develop a novel type of leukemia therapy arose during the mid-1980s, at the Shanghai Institute of Hematology.

A

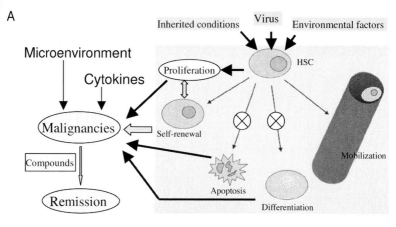

B

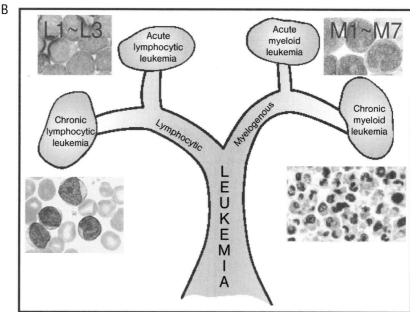

Figure 1.6 Developing targeted therapies. (A) General mechanisms for hematopoietic malignancies. (B) Leukemia tree (modified after M. Patlak).

1.3.1
Differentiation Therapy

In differentiation therapy – which in China is referred to as "educational therapy" – the approach is to educate the malignant cells to restore their normal program of differentiation and maturation, rather than to kill them (Figure 1.7).

A

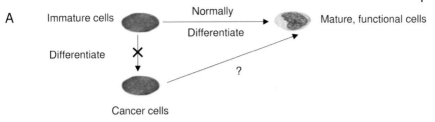

Immature cells → Normally Differentiate → Mature, functional cells

Differentiate ✗

Cancer cells

?

B

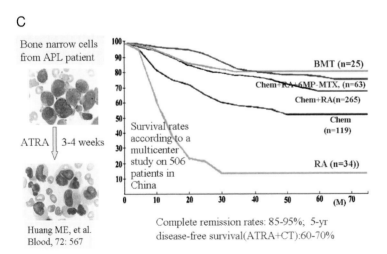

retinoids

retinol

all-trans retinoic acid

13-cis retinoic acid

9-cis retinoic acid

9,13 di-cis retinoic acid

C

Bone narrow cells from APL patient

ATRA 3-4 weeks

Huang ME, et al. Blood, 72: 567

BMT (n=25)

Chem+RA+6MP-MTX, (n=63)

Chem+RA(n=265)

Chem (n=119)

RA (n=34))

Survival rates according to a multicenter study on 506 patients in China

Complete remission rates: 85-95%; 5-yr disease-free survival(ATRA+CT):60-70%

Figure 1.7 Differentiation therapy: from hypothesis to practice.
(A) The concept of differentiation therapy.
(B) Agents used in differentiation therapy.
(C) Treatment of acute promyelocytic leukemia with retinoid combination therapy.
(After Huang et al., Blood 72, 567–572, 1988).

In this approach, the differentiation inducers used are active metabolites of vitamin A, the retinoic acids, and the disease model for differentiation therapy is acute promyelocytic leukemia (APL). It was found that one isomer of retinoic acid – all-*trans* retinoic acid – could induce the differentiation of APL cells both *in vitro* and *in vivo*, and a complete remission rate of up to 90% could be achieved using all-*trans* retinoic acid (ATRA) alone. More importantly, five-year survival rates of about 50% have been reported by groups in China and in other countries with post-remissional therapy incorporating retinoic acid and chemotherapy. It appears that, to date, these are the best results achieved for the treatment for acute leukemias in adults. Subsequently, it was demonstrated by several groups that the leukemogenic fusion protein PML-(promyelocytic leukemia) RARα (retinoic acid receptor alpha) is the result of a specific chromosome translocation t(15;17) and is the drug target for all-*trans* retinoic acid. All-*trans* retinoic acid is able to bind to the ligand-binding domain of the fusion receptor and recruit the nuclear receptor co-activator complex, thereby increasing transcription activation of the retinoic acid target genes necessary for granulocytic differentiation (Figure 1.8). Simultaneously,

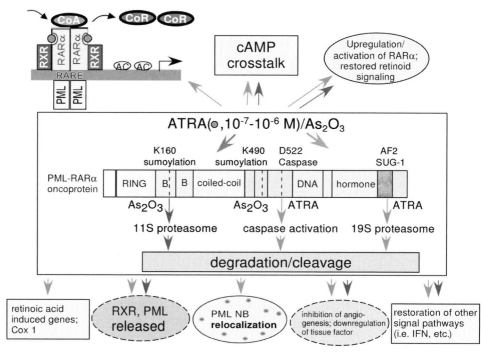

Figure 1.8 All-*trans* retinoic acid treatment for acute promyelocytic leukemia: a paradigm of targeted leukemia therapy 3.

it also induces proteasome-mediated protein degradation of the aberrant retinoic acid receptor, releasing the wild-type RARα heterodimer, as well as the PML proteins required for granulocytic differentiation and apoptosis.

1.3.2
Arsenic Trioxide

A second breakthrough in the treatment of APL was the rediscovery of an ancient drug, arsenic trioxide, as an antileukemia remedy (Figure 1.9). Arsenic exerts a dual effect on APL cells, inducing differentiation at low concentrations but apoptosis at relatively high concentrations. In fact, even in relapsed patients who have received all-trans retinoic acid and chemotherapy, arsenic can achieve a complete remission rate of 80%, which suggests that it has a unique mechanism of action (Figure 1.10).

It was subsequently discovered that arsenic could modulate and induce degradation of the oncoprotein PML-RARα, with the mode of action being quite different from that of retinoic acid. Others later showed that arsenic targets the PML moiety of the fusion receptor, with degradation of the PML-RARα being mediated through sumoylation of the PML. Based on these findings, it was proposed five years ago that all-*trans* retinoic acid and arsenic might have a synergistic effect in APL as they target the same key protein in leukemogenesis, albeit via distinct mechanisms. Indeed, at the cellular level these two agents have strong synergy in inducing degradation of the PML-RARα protein, as compared to all-*trans* retinoic acid (ATRA) or arsenic monotherapy. More importantly, the use of both drugs during remission induction and post-remissional therapy has yielded a much better molecular and a disease-free survival than in patients treated with a single agent (Figure 1.11).

This was, in fact, the result published a year ago, when no relapse was observed with a median follow-up of 18 months in a group of 20 patients treated with combination therapy as compared to monotherapy groups, where relapse occurred in a significant proportion of the patients. Whilst in biomedicine 100% relapse in not achievable, the most recent data showed that, among 45 patients with a median follow-up of 30 months, only two relapsed, and one of these subsequently responded to combination therapy. Thus, both the overall survival and disease-free survival rates were significantly better than the monotherapy groups in a comparable historic control group, and consequently it is believed that, for most APL patients, there is now hope of an effective cure.

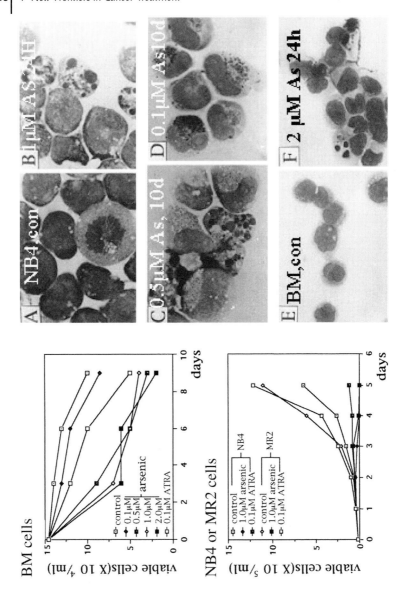

Figure 1.9 Arsenic trioxide as treatment for acute promyelocytic leukemia. (Source: Chen et al., Blood 89, 3345–3353, 1997).

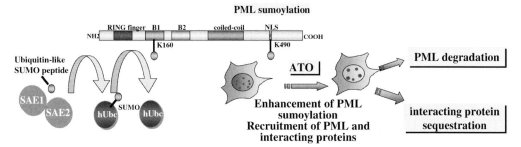

Figure 1.10 Mechanism of action of arsenic trioxide (ATO) on promyelocytic leukemia (PML) protein.

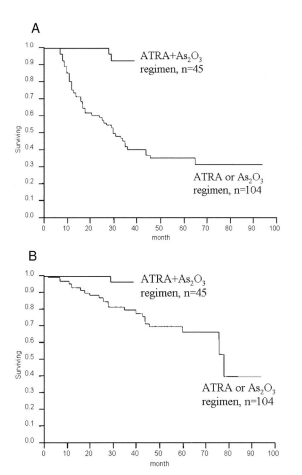

Figure 1.11 Survival rates after acute promyelocytic leukemia (APL) combination therapy. (A) Relapse-free survival. (B) Overall survival.

1.3.3
Imatinib

Another excellent example of target-based therapy is that of imatinib (Glivec, Gleevec®) in the treatment of chronic myeloid leukemia (CML). This condition is characterized by the presence of the Philadelphia chromosome which generates the PCR-ABL oncoprotein with abnormally increased protein tyrosine kinase (PTK) activity. Imatinib is able to bind in the ATP-binding pocket in the kinase domain of the ABL protein, and thus inhibits PTK activity (Figure 1.12).

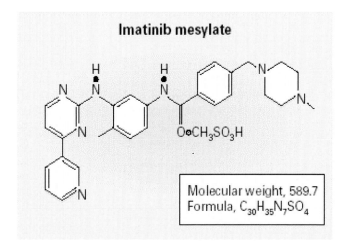

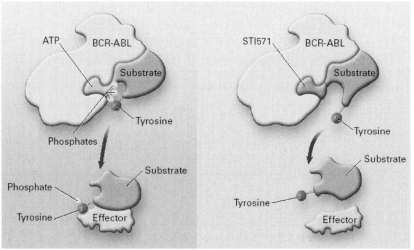

Figure 1.12 Imatinib: a protein kinase inhibitor in the treatment of chronic myeloid leukemia.

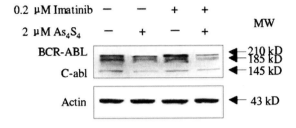

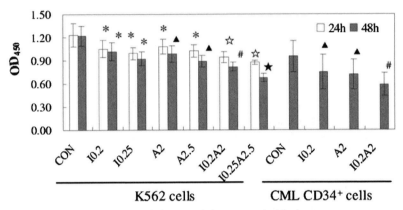

Figure 1.13 Synergistic targeted therapy of chronic myeloid leukemia with arsenic (A) and imatinib (I).

Today, imatinib is the new "gold standard" in CML therapy, and provides durable clinical cytogenetic and molecular remission in CML patients, particularly those in the chronic phase. However, with long-term use of imatinib, some patients develop resistance to the drug. According to investigations with APL, it is believed that a combination therapy might be more effective than single-agent therapy, even in the case of imatinib. A review of the literature showed that one drug which might, potentially, be combined with imatinib is arsenic sulfide. This differs from arsenic trioxide in that it exhibits a synergistic effect by inducing cell apoptosis with imatinib (Figure 1.13). The apoptosis is much more profound in cells treated with both agents than in those treated with a single drug. It was subsequently found that arsenic sulfide could significantly reduce the level of BCR-ABL protein (see Figure 1.13, upper panel), whereas imatinib inhibits the PTK activity of the oncoprotein. Interestingly, when the two drugs operate together, the reductions in BCR-ABL oncoprotein level and PTK activity are much more significant than with monotherapy. There is therefore, another paradigm of synergistic targeting therapy, not on a transcription factor such as PML-RARα but on a signaling molecule, the BCR-ABL. The results obtained in clinical trials using combination therapy have so far been intriguing, however.

The mechanism of leukemogenesis remains elusive, however. In acute myeloid leukemia (AML) with major chromosome abnormalities, the molecular mechanisms have been well characterized, but in AML without identifiable chromosome abnormalities very little is still known about these mechanisms. The same is true for the acute transformation of CML. In order to further dissect the mechanisms of leukemogenesis and to open new therapeutic perspectives, the Shanghai Institute of Hematology recently launched the Leukemia Genome Anatomy Project (LGAP) to survey abnormalities in genes critical to the regulation of hematopoiesis and onset of leukemia. This project is accompanied by the Leukemia Integrative Chemical Genomics Project, which aims to facilitate target validation and compound screening. In fact, this is mostly based on the screening of natural compounds from traditional Chinese medicine.

Although surgery remains the most important treatment for patients with solid tumors, targeted therapy now plays an increasingly important role in this respect. The first such example is apparent in breast cancer, where expression of the oncogene HER-2 can be detected in 20–25% of cases, while the expression of HER-2 correlates with poor disease-free survival and resistance to chemotherapy and endocrine therapy. An antibody known as trastuzumab (Herceptin), and a humanized mouse anti-HER-2 monoclonal antibody, were found to inhibit HER signaling and thereby inhibit disease progression and enhance survival (Figure 1.14). Today, Herceptin plus chemotherapy represents the best drug treatment for metastatic breast cancer.

Currently, many other anti-cancer antibodies are available for the therapy of diseases such as solid tumor, lymphoma, and leukemia (Table 1.2). It has been well established that sustained angiogenesis is crucial to cancer, and cancers can by themselves trigger the growth of new vessels by inducing the secretion of growth factors (e.g., vascular endothelial growth factor; VEGF). Moreover, the angiogenesis may support tumor growth and metastasis by supplying nutrition to the cancer cells. Anti-angiogenesis is therefore an evidence-based therapeutic strategy. Approaches to modulate VEGF or VEGF receptor signaling include the targeting of VEGF or its receptor with antibodies that reduce VEGF expression by using ribozymes and RNA interference (Figure 1.15). Anti-angiogenesis, as expected, may cause vessels to regress, and this would lead to tumor shrinkage. This anti-cancer strategy has now achieved some degree of clinical success in metastatic colorectal cancer, it being reported recently that the anti-VEGF antibody, bevacizumab, increased overall survival when co-administered with chemotherapy as compared with chemotherapy alone (Figure 1.16).

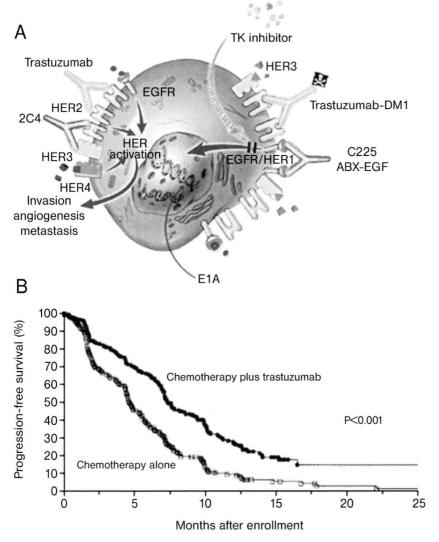

Figure 1.14 Trastuzumab as an example of an anticancer antibody therapeutic.
(A) Areas of potential therapeutic intervention.
(B) Trastuzumab combination therapy of breast cancer.

Table 1.2 Overview of FDA-approved targeted anticancer antibody therapeutics. (Source: Ross et al., Am. J. Clin. Pathol. 122, 598–609, 2004).

Name	Date approved	Source (partners)	Type	Target	Approved indication	Diagnostic test required
Alemtuzumab (Campath)	May 2001	ILEX Oncology, San Antonio, TX; Schering, AG, Berlin, Germany	Monoclonal antibody, humanized; anticancer, immunologic; multiple sclerosis treatment; immunosuppressant	CD52	CLL	no
Rituximab (Rituxan)	November 1997	IDEC, La Jolla, CA (Genentech, South San Francisco, CA; Hoffmann-La Roche, Basel, Switzerland; Zenyaku Kogyo, Tokyo, Japan)	Monoclonal IgG1; chimeric; anticancer, immunologic; antiarthritic, immunologic; immunosuppressant	CD20	NHL	no
Trastuzumab (Herceptin)	September 1998	Genentech (Hoffmann-La Roche; ImmunoGen, Cambridge, MA)	Monoclonal IgG1 humanized; anticancer, immunologic	p185neu	Breast cancer	yes
Gemtuzumab (Mylotarg)	May 2000	Wyeth/AHP, Madison, NJ	Monoclonal IgG4 humanized	CD33/calicheamicin	AML (patients > 60 y)	no
Ibritumomab (Zevalin)	February 2002	IDEC	Monoclonal IgG1 murine; anticancer	CD20/^{90}Y	NHL	no
Edrecolomab (Panorex)	January 1995	GlaxoSmithKline, London, England	Monoclonal IgG2A murine; anticancer	EpCAM	Colorectal cancer	no
Tositumomab (Bexxar)	June 2003	Corixa, Seattle, WA	Anti-CD20 murine monoclonal antibody with ^{131}I conjugation	CD20	NHL	no
Cetuximab (Erbitux)	February 2004	Imclone, New York, NY; Bristol Myers, Princeton, NJ	Anti-EGFR	EGFR	CRC in combination with CPT-11 (irinotecan)	yes
Bevacizumab (Avastin)	February 2004	Genentech	Anti-VEGF (ligand)	VEGF	CRC; first line in combination with 5-FU	no

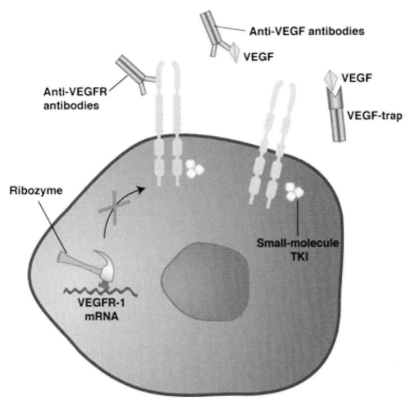

Figure 1.15 Approaches to modulate VEGF/VEGFR signaling.
(Source: Steward, Horizons in Cancer Therapeutics 5, 11–21, 2004).

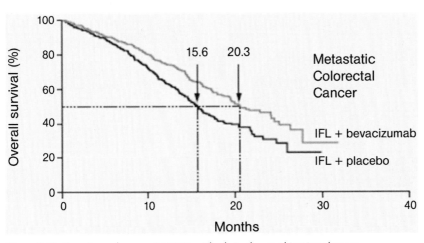

Figure 1.16 Bevacizumab: an anti-VEGF antibody in the combination therapy
of colorectal cancer (IFL = irinotecan, fluorouracil, leucovorin).

1.4
Lung Cancer

The question of how the clinical outcome of lung cancer treatment might be further improved is urgent, notably because among cancer patients this condition is the number one killer worldwide, including China. Non-small-cell lung cancer (NSCLC) accounts for about 85% of all lung cancers, and a major advance during the past few years has been the use of a new small compound, gefitinib, in the treatment of this condition. Gefitinib treatment induces a marked remission of lung cancer in about 10% of patients in western countries and in 30% of cases in oriental countries, particularly in Japan and China (Figure 1.17).

The mode of action of gefitinib in the treatment of lung cancer is unclear, although inhibition of the mutant epidermal growth factor receptor (EGF receptor) is known to underlie gefitinib's anti-lung cancer activity. In contrast, gain-of-function mutations of the EGF receptor are critical to carcinogenesis of NSCLC. In ancient China, one of the main principles of combating disease was to treat a disease before its onset, and this might represent the advent of preventive medicine in oriental countries. The prevention of cancer from its onset, or of cancer progression, may be seen in many forms. For example, tobacco smoking can cause cancer of the lung and of other organs, as well as other respiratory and/or cardiovascular conditions, but when the patient stops smoking the improvements are remarkable. Typically, in a study of 14.5 years' duration, the mortality rate due to lung cancer was much lower in sustained "quitters" of smoking than in those who quit intermittently or who continued to smoke (Figure 1.18).

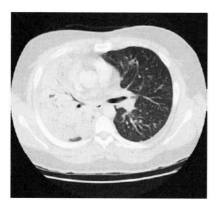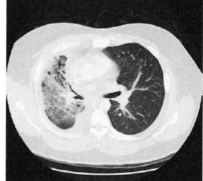

Figure 1.17 Gefitinib in the therapy for non-small cell lung cancer. Tomographic scans of a patient with lung cancer before (left) and 6 weeks after treatment with gefitinib (right). (Source: Lynch et al., N. Engl. J. Med. 350, 2129–2139, 2004).

1.5
Outlook

In China, the widespread use of hepatitis B vaccination dramatically reduced the virus infection ratio, and consequently the number of patients suffering from liver cancer has decreased by 25% over the past 15 years. It can be concluded, therefore, that a skillful doctor cures illness when there is no sign of disease, and thus the disease never materializes – as was proposed in China by Dr. Huai Nan Zi, some 2100 years ago.

The main steps for cancer prevention, though not easy to maintain, are clear – to stop smoking and to avoid alcohol over-indulgence, to eat a healthy diet, and to protect the body against sunlight, X-rays, chemicals, industrial agents and viruses. In addition, it is advisable to maintain a healthy body-weight, to stay active, and to undergo routine cancer screening. Knowledge concerning certain medicines that might prevent cancer, for example green tea, might also be beneficial.

In conclusion, cancer represents a group of heterogeneous malignancies and genetic abnormalities that underlie the elusive process of carcinogenesis. At present. molecular target-based therapies show great promise and, when combined with conventional procedures, may indeed represent the future of cancer treatment. Thomas Edison once said, "I am long on ideas, but short on time, I expect to live only 100 years". Of course, everybody – including

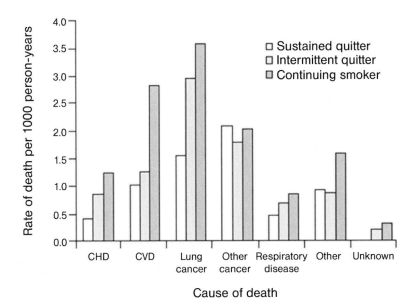

Figure 1.18 Influence of smoking habits on lung cancer.
CHD = coronary heart disease; CVD = cardiovascular disease.

those suffering from cancer – wishes to live longer, and winning the war against cancer forms the focal point for those investigating the therapy and research of the condition. However, this cannot be achieved without the support of the general public, of the patients, of the pharmaceutical industry, of the decision makers, and of society in general.

Author Biography

Ciro A. de Quadros

Director, International Programs, Sabin Vaccine Institute (US)

Ciro de Quadros, M.D., M.P.H., has dedicated his career to freeing the world of infectious diseases, especially those that disproportionately affect the health and social development of the world's poorer countries.

At present he is the Director for International Programs at the Sabin Vaccine Institute (SVI), since February 2003. Before joining the SVI he was Director of the Division of Vaccines and Immunization of the Pan American Health Organization in Washington, D.C.

He completed his medical studies in Brazil in 1966 and received a Master of Public Health degree from the National School of Public Health in Rio de Janeiro in1968. He was involved with the pioneering experiences for the development of the strategies of surveillance and containment for Smallpox eradication and in February, 1970 joined the World Health Organization (WHO) as Chief Epidemiologist for the Smallpox Eradication Program in Ethiopia. He transferred to the Pan American Health Organization (PAHO) in February 1997 to serve as the Senior Advisor on Immunizations. While working at PAHO he directed the successful efforts of polio and measles eradication from the Western Hemisphere. He retired from PAHO in 2002 and since 2003 has led Albert B. Sabin Vaccine Institute's international programs, with emphasis in advocacy activities directed at the control and/or elimination of rotavirus and rubella in Latin America and the Caribbean. He is also the chairperson of the Independent Review Committee of the Global Alliance for Vaccines and Immunization (GAVI) and serves on the Board of the International Aids Vaccine Initiative (IAVI). He is on faculty at the Johns Hopkins Bloomberg School of Hygiene and Public Health and the Schools of Medicine at Case Western Reserve University, in Cleveland and the George Washington University, in Washington, DC. He has published over 120 scientific papers and book chapters in peer reviewed journals and has lectured in over 200 scientific meetings and conferences throughout the world. Ciro A. de Quadros has received several international awards, including the 1993 Prince Mahidol Award of Thailand, the 2000 Albert B. Sabin Gold Medal, the Order of Rio Branco from his native Brazil, and most recently, was named International Health Hero by the School of Public Health of the University of California at Berkeley. In 2004 he was elected Member of the Institute of Medicine of the National Academies of the United States of America.

2
Eradication of Vaccine-Preventable Diseases: Is it Possible and Cost-Beneficial?

Ciro A. de Quadros

2.1
Introduction

With regard to the question of whether investments in new therapies are paying off, there is perhaps no better example than the role of vaccines and their power in controlling diseases. Previously, much has been written about vaccines being the most effective health intervention system currently available, and this chapter will focus on some aspects and examples of disease eradication through vaccine use. This will include examples of successes and failures and aspects regarding the organization of eradication programs, followed by a brief description of the cost-effect and cost-benefit of the eradication of poliomyelitis and measles in the western hemisphere. The chapter will conclude with some possible insights into the future of disease eradication, and outline some of the lessons that have been learned so far.

2.2
Disease Eradication

By definition, disease eradication is the absence of a disease agent in nature in a defined geographical area, whereafter control measures can be discontinued once the risk of importation is no longer present. Whether control measures are interrupted or not, a disease can still be eradicated. Such a concept is permanently evolving in terms of eradication, and this can be seen from the definitions proposed at a meeting held at Dahlem near Berlin, Germany, several years ago, when "eradication" was defined in terms of different levels of disease control (Table 2.1, upper portion). On the first level, the disease is reduced to acceptable levels, and then proceeds to a level defined as "elimination". Continuing, there is the elimination of infection in a defined

Health for All?: Analyses and Recommendations
Edited by The World Life Sciences Forum – BioVision
Copyright © 2005 Wiley-VCH Verlag GmbH & Co. KGaA, Weinheim
ISBN: 3-527-31489-X

Table 2.1 Levels of control of vaccine preventable diseases.

Dahlem definitions (1997):

- Control = reduction to acceptable levels
- Elimination (of disease or infection) = no new cases in defined areas; control measures are maintained
- Eradication = worldwide absence of new infections; control measures can be discontinued
- Extinction = infectious agent no longer exists in nature and laboratory

---

Post-Dahlem (Decatur) definitions:

- Control = reduction to acceptable levels
- Eradication = no new infections in defined areas; control measures can be discontinued if there is no risk of reintroduction
- Extinction = infectious agent no longer exists in nature and laboratory

---

area, again with continued measures, until eradication and finally elimination at the global level is reached. At this point, the measures may be discontinued. The final stage is extinction of the disease, which refers to the elimination of an infectious agent, both in nature and in the laboratory. These definitions have been the subject of much debate, since many of the terminologies translate differently in different languages. Thus, a smaller group met in Decatur, Georgia, to define control as simply three levels: (1) a reduction to acceptable levels; (2) eradication in a defined area, which might be regional, worldwide and control measures which could be discontinued if no risk of introduction existed; and (3) extinction, both in nature and in the laboratory (Table 2.1, lower portion).

2.2.1
Criteria of Disease Eradication

In order to eradicate a disease it is necessary to fulfill certain criteria which are basically biological and/or societal. In this respect, humans must be essential to the life cycle of an infectious agent, and there is a specific intervention to interrupt transmission of the infectious agent from one person to another. In practice, diagnostic instruments are available that will be sufficiently sensitive and specific to detect transmission of the infectious agent. Finally, there is a societal interest in the eradication of a disease, and this of course is related to the disease burden – how the disease is seen by society and the cost to eradicate it.

Among previous eradication initiatives, perhaps the most well-known (and first) viral disease to be eradicated was smallpox. Edward Jenner had declared more than 200 years ago that the use of his vaccine would extinguish smallpox

Table 2.2 Past eradication initiatives.

Year	Initiator	Disease	Region	Outcome
1801	Jenner	Smallpox	Global	Failed
1911	Gorgas	Yellow fever	The Americas	Failed
1915	Rockefeller Commission	Yellow fever	Global	Failed
1950	Soper	Smallpox	The Americas	Failed
1955	WHO	Malaria	Global	Failed
1958	Zhdanov	Smallpox	Global	Succeeded
1985	PAHO	Polio	The Americas	Succeeded
1988	WHO	Polio	Global	Uncertain
1994	PAHO	Measles	The Americas	Succeeded
2003	PAHO	Rubella	The Americas	Uncertain

from the face of the earth (Table 2.2). In 1904, King Charles III of Spain sent an expedition to the Spanish colonies in America and the east in an attempt to stop the transmission of smallpox. In 1911, Gorgas attempted to eradicate yellow fever in the Americas, and this was followed by the Rockefeller Commission's attempt to globally eradicate this disease. In 1950, the Pan-American Health Organization (PAHO) launched an effort to eradicate small-pox in the Americas. The World Health Organization (WHO) attempted to eradicate yaws on a regional basis, and also tried to eradicate malaria globally. In 1958, Professor Zhdanov from the then Soviet Union proposed the eradication of smallpox, whereafter PAHO in 1985 proposed the eradication of polio in the Americas. In 1986, the WHO started a program for Guinea worm eradication in most parts of Africa, while in 1988, following the success of the eradication of polio in the Americas, the WHO launched a polio eradication scheme on a global level. In 1994, when the Americas were certified polio-free, the PAHO launched an effort to eradicate measles and in 2003, when the transmission of measles had been interrupted in the Americas, the directing council of the PAHO again launched a program to eradicate rubella from the Americas.

However, most of those initiatives, including yellow fever and malaria, failed because they did not fulfill the biological criteria for eradication. In fact, to date only the global effort to eradicate smallpox has succeeded, though it is hoped that polio eradication will in time also be successful. Measles eradication has been successful in the Americas, but this has not quite been the case for rubella, although progress to date has been outstanding. However, it is believed that during the next two to three years, rubella transmission will be interrupted in the region of the Americas.

2.3
Launching Eradication Programs

The next question is how and when to launch eradication programs. Clearly, it is essential to examine the reality of science – the biological versus the hope factor – since the result can be fatal if these aspects are not considered. An example of this is the failure of malaria eradication, which almost led to the breakdown of the WHO.

One very important factor is the duration of the program, and in this respect epidemiological models allow committed partners to share the financial responsibilities. The tension that exists from a horizontal approach versus a vertical approach must also be recognized. This debate has raged for more than a century, since it relates to the importance of the strategy that will be applied to the eradication program. Certain other factors that will influence the coverage of a program must also be considered, including the technicians involved, the feasibility of the study, and the adaptation and issue of human resources, which is critical. The perception of the problem by politicians, the trust of the community, and sometimes even the coercion of the community (as occurred in many regions of the world during the smallpox eradication program), and finally the commitment of funding, costs and economics involved, will all impact upon the duration of the program. Put simply, an eradication program cannot continue for ever – it must have only a limited (usually very limited) lifetime.

A number of external factors must also be considered, some of which are positive and some negative. The former category includes a guarantee of the common principles that create mutual benefits, in particular the use of eradication programs to reinforce health systems and infrastructure in different countries. An important example of this was poliomyelitis in the Americas, since the program was launched not only to eradicate polio but also to reinforce the infrastructure of health systems, and particularly the surveillance and logistical systems.

Negative aspects include a suboptimal implementation that may lead to resistance, thereby increasing the long-term problem, and economic difficulties such as the development of human resources, the laboratory network that can be created (and has been created with polio eradication and measles eradication in the Americas), and the political capital for public health. In developing countries, the weakest minister is usually the one holding the health portfolio, and conquering or re-controlling disease will increase the political capital of that sector. These negative aspects may become dangerous practices if left uncared for, and increased costs may lead to the termination of coordinated schemes.

2.4
Examples of Disease Eradication

The two main examples of eradication from the Americas are those of polio and measles.

2.4.1
Polio

The strategy used initially for polio – and now utilized worldwide – was the routine immunization of children by the health services, coupled with supplemental immunization combined with national immunization days and mopping-up operations with surveillance of acute flaccid paralysis. This approach was first implemented in Cuba, when Albert Sabin first developed the oral polio vaccine and showed that national immunization days could stop transmission; indeed, Cuba was the first country to stop the transmission of polio in 1962. As a result, this approach was implemented in all countries in the Americas, such that the disease disappeared (Figure 2.1). The last indigenous case of polio occurred in 1991 in Peru, and in 1994 the region was declared polio-free. In 2000, there was a resurgence in the Hispaniola Island, with an outbreak provoked by a reversion of the vaccine virus.

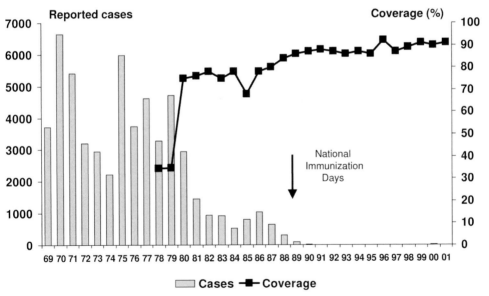

Figure 2.1 OPV3 vaccination coverage and incidence of paralytic poliomyelitis in the Americas. Coverage is for children aged over 1 year. Data for 2000 and 2001 represent Type 1 vaccine derived virus. (Source: HVP/PAHO).

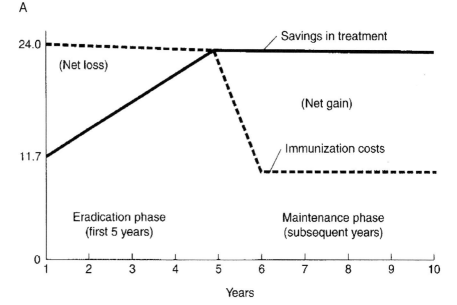

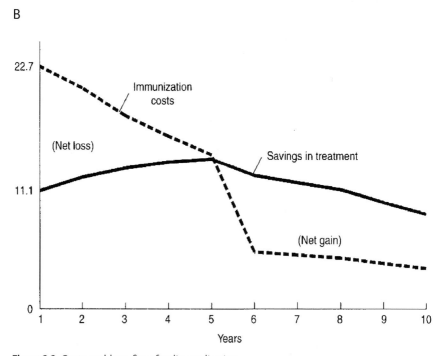

Figure 2.2 Costs and benefits of polio eradication.
(A) No discounting. (B) 12% discounting per year.
The study assumes treatment of only some victims.
(Source: P. Musgrove, Health Economics and Development, World Bank, 2003).

Table 2.3 Costs and benefits associated with polio eradication during a
successful five-year campaign and an ensuing 10-year maintenance period.
(Source: P. Musgrove, Health Economics and Development, World Bank, 2003).

Total costs versus benefits (figures are US$ million, except where indicated)	Treatment of all victims		Treatment of only a portion of the victims	
	First 5 years	All 15 years	First 5 years	All 15 years
No. of cases prevented ('000)	70	220	15	55
Savings in treatment expenses	408.0	1282.4	87.5	320.7
Cost of eradication or maintenance	120.0	220.0	120.0	220.0
Net savings (net benefit)	288.0	1062.4	−32.5	100.7
Net present value of discounted savings	217.2	481.4	−27.3	18.1

The costs and benefits associated with polio eradication during the successful
five-year campaign and an ensuing ten-year maintenance period are listed in
Table 2.3 and illustrated graphically in Figure 2.2. The net present value of
discounted savings was US$ 480 million for the period under study, and even
if all cases had not been treated (as was likely in many developing countries
and Latin America), the net present value of discounted savings was still in
the order of US$ 18 million.

The ultimate justification for a US$ 2 billion polio eradication scheme is
that the children involved would grow up to be healthy, and would be
productive members of society, and thereby contribute to the GMP of their
country. Clearly, consideration must be given to human lives and suffering
rather than simply to the dollars involved.

2.4.2
Measles

The measles elimination strategy in the Americas relied on the catch-up
campaign – that is, the vaccination of all children aged between 1 and 14
years with one dose of measles vaccine during the low season, the aim being
to interrupt transmission. This was followed up by vaccination during routine
services (termed the "keep-up"), and a very high coverage was achieved in
children aged 12 to 23 months in order to maintain an interruption of
transmission. Obviously, this is possible if the health systems can be reinforced
so that they deliver vaccines on a daily basis. It is important that the maximum
coverage possible is achieved throughout the country, and here again vaccines
may represent one of the most equitable health interventions, simply because
they can reach every child, as has been shown in several efforts worldwide.

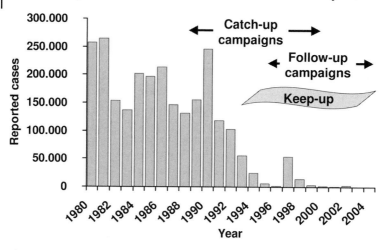

Figure 2.3 Measles cases in the Americas, 1980–2004.

Periodic mass campaigns, termed "follow-up" campaigns, are conducted every four years, and are aimed at vaccinating children aged 1 to 4 years. The goal would be to provide the first dose to those children who missed their first dose during the routine service. This approach is now referred to by WHO and UNICEF as a second opportunity for the child to obtain their first dose of vaccine, though of course the majority will receive a bonus second dose, thereby maintaining an interrupted transmission. Figure 2.3 shows, graphically, the impact of catch-up campaigns on measles cases in the region. The campaigns were again started in Cuba, followed by the English-Caribbean and other countries, such that over a period of about 8–10 years, all of these countries had implemented campaigns and improved their routine services, after which periodic follow-up campaigns were implemented.

2.5
The Costs of Immunization

A breakdown of the components of routine immunization programs is illustrated in Figure 2.4. The highest expenses are related to administrative costs and vaccine delivery. When examining the aggregate cost of measles elimination in Latin America, the savings over a 21-year period were on the order of US$ 200 million (Table 2.4). Again, the dollar figures were high, but more important is the fact that over the past few years, no child has died of measles in Latin America, or in the whole of the western hemisphere.

Sadly, in Japan about two or three years ago, over 50 children died from measles. It is unacceptable that in today's world, where vaccines cost so very

A

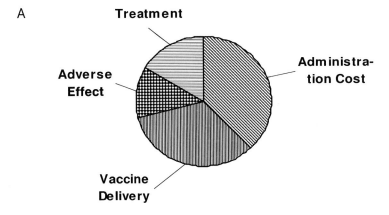

B

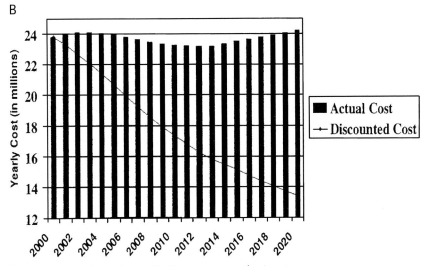

Figure 2.4 Cost of measles eradication program in the Americas.
(A) Component breakdown of yearly costs for 2000.
(B) Estimated yearly costs for 2000–2020.
(Source: Acharya et al., Vaccine 20, 3332–3341, 2002).

little, children are still allowed to be paralyzed by polio, to die from measles, or be born with malformations due to congenital rubella. Society must change this social "norm".

On the other hand, the cost of failure and the loss of trust in public health programs can be enormous. Consequently, extreme care must be taken when the decision is taken to eradicate a disease. There is also a loss of opportunities to introduce new vaccines, as well as biological issues related to pressure in the selection of vectors, or resistance to parasites that will increase the cost of control in the future.

Table 2.4 Aggregate cost of measles elimination in Latin American countries. All costs in the lower half of the table are discounted at 3% and extrapolated to entire Latin America. (Source: Acharya et al., Vaccine 20, 3332–3341, 2002).

Description	Cost (US$ million, 1999)
Cost of routine program	
• Year 2000	23.8
• Years 2000–2020 (discounted at 3%)	376.4
Cost of follow-up program	
• Year 2000	38.4
• Years 2000–2020	172.0
Cost of the entire program	
• Years 2000–2020	548.4
• Extrapolated to entire Latin America	571.2
Cost of measles program without elimination effort	779.5
Cost savings due to the elimination program	
• Years 2000–2020	208.3

In this respect, political commitment is critical, and strategies must be clearly understood at all levels of the health system, as well as by those involved in the implementation. Clearly, it is not possible to have a rigid strategy, and research data must be available that fit into the program, so that a strategy can be adapted as it progresses. The resources must also be adequate and easily available. For example, when the program was launched in the Americas, the resources were not easily obtained, and indeed a chronic problem of global polio eradication is that the resources are always in short measure. There must also be strong management, research facilities to guide the strategy, adequate international coordination, a motivated staff, and a time limit for the completion of the program.

2.6
Future Prospects

When considering future prospects for the eradication of viral diseases, a number of approaches have been made in the past, and clearly more will be made in the future. Some of the diseases listed in Table 2.5 are not eradicable, some form the basis of a major disease burden worldwide, and in societal terms some are not even on the politicians' "radar screens". However, if polio can be eradicated, then measles may well be the next target for eradication.

Table 2.5 Prospects for the eradication of viral diseases.

Disease	Eradication feasible?	Eradication likely?
Yellow fever	no	–
Rabies	no	–
Juvenile encephalitis	no	–
Influenza	no	–
Varicella	no	–
Hepatitis A	yes	no
Hepatitis B	yes	no
Mumps	yes	no
Rubella	yes	yes
Measles	yes	yes

Likewise, rubella eradication is succeeding in Latin America and in the western hemisphere, and in March 2005 it was declared as being "interrupted" in the United States.

2.7
Summary

In conclusion, it is important to understand the natural history of a disease before launching a program for its eradication. There must be extensive consultation before program initiation, and surveillance must be carried out from the start of the program in order to provide the strategy with guidance. The approach also requires a vertical tactic which, if well applied, may be very helpful for improving health systems. Moreover, it is necessary to expect the unexpected, so there must be flexibility. Of course some countries will require more assistance than others, and this situation must be well analyzed, the costs considered, and the funding readily available. In this respect, the coordination of partners is essential, and victory must not be declared too soon. It was Louis Pasteur who stated that, "... it is within the power of man to eradicate infection from the earth", and there is no other health intervention more powerful than vaccines.

Author Biography

Peter B. Corr

Senior Vice President, Science and Technology, Pfizer

Dr. Peter B. Corr is Senior Vice President for Science and Technology at Pfizer Inc., where he is responsible for aligning the company's worldwide research and development organization with licensing activities, science and medical advocacy, global medical relations and science policy.

He has had a long and distinguished career in science and technology, including positions in both academia and industry. Prior to joining Pfizer in 2000, he held senior leadership positions in research, development and discovery at both Warner-Lambert Company and Monsanto/Searle.

Peter B. Corr, who received his PhD from Georgetown University School of Medicine and Dentistry, spent more than 18 years as a leading researcher in molecular biology and pharmacology at Washington University. His research, published in more than 160 scientific manuscripts, has focused on the biochemical mechanisms, with particular emphasis on ischemic heart disease, sudden cardiac death and other abnormal cardiac rhythms.

He is the recipient of numerous awards, including Alpha Omega Alpha National Medical Honorary Society, an Established Investigator Award from the American Heart Association, and a Research Career Development Award from the NIH. In 1981, he received the Washington University School of Medicine Teacher of the Year Award, and in 1980 the Washington University Distinguished Faculty Award.

He has also served on or is serving on the editorial board of several journals, including *American Journal of Physiology, Circulation, Circulation Research, Molecular and Cellular Biochemistry, and the Journal of Cardiovascular Electro-physiology.*

3
Incentives in Policy Reforms Necessary to Stimulate Activity

Peter B. Corr

3.1
Introduction

In my contribution I would like to emphasize both the challenge and opportunity to make clear improvements in the difficult health problems that we are facing today, both in the developed and developing worlds. It is possible that we are victims of our own success, as science has become such a massively complex body of knowledge that no one person, discipline, institution or company can manage it in its entirety. The technological demands and complexity of today's research methods are not only "mind-boggling", they are also very expensive to conduct. At the same time, from a policy perspective, the array of regulations that medicines face as they spread across the globe creates another set of challenges. Medicines are the most regulated entity in the world, and indeed it is only right that they should be. They must show both efficacy and safety. To maintain this requires a concerted effort to be made by all concerned, but we also have to think about how we balance the risks and benefits.

3.2
High Attrition: High Costs

Today, one of the key challenges faced by the pharmaceutical industry is that of high failure rates and attrition in research and development as, taken together, these rapidly drive up costs. In general, a basic research finding – that is, a particular target or targets – leads to the development, via biological and chemical routes, of a particular molecule that will interact specifically with that target (Figure 3.1). Subsequently, a single compound is selected and moved into preclinical development, where it undergoes very expensive testing

Health for All?: Analyses and Recommendations
Edited by The World Life Sciences Forum – BioVision
Copyright © 2005 Wiley-VCH Verlag GmbH & Co. KGaA, Weinheim
ISBN: 3-527-31489-X

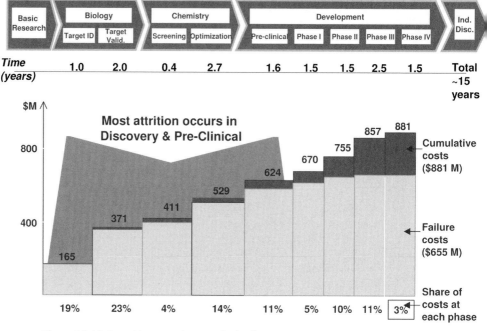

Figure 3.1 High attrition contributes to high R&D costs.

in areas of toxicology and pharmacokinetics. Finally, the molecule is moved into man where, with good fortune, the new drug passes through various stages of clinical development and then to regulatory approval. The overall duration of this process varies from drug to drug, but typically ranges from 10 to 15 years.

Why is this process so expensive? The analysis shown in Figure 3.1, which was conducted by examining data from several companies, suggests that a figure of US$ 880 million is reasonable. However, by examining the share of cost at each phase, it is clear that the main problem is that of failure costs which, in the case illustrated, is US$ 655 million on average. Of course, if in some way success could be identified at the start of the process, this would reduce costs by some 75%. Attrition is very high during the preclinical phase. In consequence, millions of compounds are screened by the pharmaceutical companies each year, yet very few reach the marketplace.

The world as a whole seems to believe that much more is known about the biology of disease than is true. In this respect the learning process is continuous, and much valuable information is obtained from the failures. The result is that, despite these enormous challenges, there is great optimism within the pharmaceutical industry that the cusp of a new paradigm is rapidly approaching.

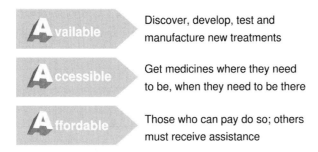

Discover, develop, test and
manufacture new treatments

Get medicines where they need
to be, when they need to be there

Those who can pay do so; others
must receive assistance

Figure 3.2 The three "A"s of the biomedical industry.

3.3
Availability, Accessibility, Affordability

At Pfizer, the commitment is to making medicines available, accessible, and affordable – what is known as the three "As". Although effective, this strategy and commitment requires both good science and good business (Figure 3.2):

- First, medicines must be made *available* – that is, they must be discovered, developed, manufactured and distributed.

- Second, they must be made *accessible*, physicians and patients must be taught about them, and they must be dispatched where and when they are needed, worldwide. Accessibility has other connotations, however, in terms of infrastructure, delivery systems, and health professionals working across sectors to connect the patient to the therapy, and this represents a major problem in the developing world.

- Finally, medicines must be made *affordable*. The challenge is to ensure that third-parties pay for medicines when appropriate, and to provide help for those people who cannot pay for them by their own means when necessary, and to price according to the markets in which they are accessible.

This entire system rests on the foundation of the availability of new medicines. Availability is a matter of science and large investment.

Today, more than ever, there are three indispensable elements that help make these medicines available, accessible and affordable:

- The unique abilities of the biomedical industry to discover, test and develop new therapies on a sufficiently large scale, and with efficiencies sufficient to meet the global burden of disease are essential.

- The involvement of the public sector is essential to maximize the efforts of the biomedical community by advancing basic science in both the private and public sectors. It is also necessary to identify how better regulatory strategies and efficiencies can be provided, to support a sound intellectual property system, and to use such a system to provide incentives in the right way. It is also important to balance the need with a careful understanding of what is scientifically possible.

- There must be stronger partnerships across all sectors, in order to prevent duplication and to maximize efficiencies. Partnerships represent the real goal, as the complexities, risks and high costs of R&D demand that members of the biomedical community coordinate their efforts. Much more can be accomplished, with much less redundancy, if institutions and organizations work together and bring their combined strengths to bear.

3.4
Innovation and Partnership

In the spirit of partnership, what is needed to fuel innovation and to realize the promise of biomedical endeavor? Although tens of billions of dollars are being spent, the opportunity remains largely unrealized. The requirements are clear. First and foremost is a successful healthcare system that provides an efficient delivery and distribution of services, together with efficient pricing and reimbursement for these services. Healthcare systems must be sustained by the effective use of intellectual property, including the enforcement of intellecual property rights, such that investment in discovery, basic science and development will continue. There should also be prevention of parallel trade, so that patients in developing countries will have access to affordable medicines, usually at zero or near-zero cost. Healthcare systems must be supported by adequate and predictable regulatory requirements, including a safe and efficient drug approval process and, importantly, a global harmonization of regulatory requirements. This may well be the key to success, as a great deal of money is spent performing different studies for different regions of the world. A rapid adjustment of regulatory requirements to current advances in science and technology is also requisite.

How do we know these elements are so critical to innovation? Because the record proves it. During the past 20 years, the pharmaceutical and biotechnology industries have brought to patients well over 90% of all new medicines developed worldwide. In fact, today there are 700 new therapies in development alone, all at high risk to the companies and at very high cost. It is also known that innovative medicines have the capacity to reduce morbidity and to lessen the need for expensive hospital care. As an example, treatment with inexpensive cholesterol-lowering drugs can reduce the need for subsequent expensive angioplasty procedures and coronary artery bypass graft surgery. Over the past 40 years, the use of medicines has helped to reduce numbers of hospital admissions by half for 12 major diseases, including ulcers, mental illness and infections. Moreover, it is known that investments in R&D contribute to economic competitiveness, and that more competitive medicines can consistently reduce the cost burden of disease, particularly if patients are diagnosed earlier, treated earlier, and compliance to treatment is markedly improved compared to the present situation. This latter point applies not only to the developed world but also to some extent to the developing world.

3.5
The Economic Value of Health

In a study conducted by Kevin Murphy and Robert Topel, two economists at the University of Chicago, an analysis of US data showed that even modest reductions in death rates due to common killers such as cancer, heart disease, diabetes, pneumonia and influenza could produce literally trillions of dollars in added economic benefit to society via these affected individuals over their lifetime. For example, a 10% reduction in deaths from diabetes – bearing in mind the rapid worldwide increase in diabetes – would provide US$ 450 billion in overall economic benefits to society during the lifetime of America's present living population (Table 3.1). Moreover, this situation can be extrapolated throughout the world.

Table 3.1 The economic value of health.

Cause of death	Impact of 10% decrease (US$ billion)
Major cardiovascular diseases	5142
Malignant neoplasms	4359
Infectious diseases (including AIDS)	644
Chronic obstructive pulmonary diseases	605
Pneumonia and influenza	358
Diabetes	449
Chronic liver disease and cirrhosis	310
Accidents and adverse effects	1369
Homicide and legal intervention	413
Suicide	508
Other	3006
All causes	17163

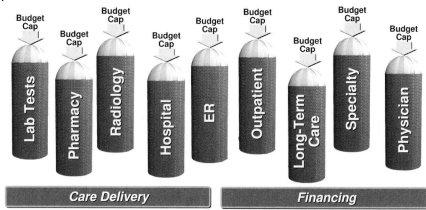

Figure 3.3 The current healthcare management and delivery paradigm.

These examples of economic benefit are exactly why healthcare regulators and governments should focus on the cost burden of disease than simply on the cost of therapies to treat that disease. Unfortunately, the problem in Europe and the US – and to some extent in Japan and other countries in the developed world – is that the approach towards healthcare resembles a series of many unconnected "silos" (Figure 3.3). Whether it is the hospital, the physician, the pharmacy, or even the insurer, each silo only considers ways to reduce costs within its own sphere of influence. Yet if healthcare is considered in its entirety, from preventive measures to early diagnosis and treatment, it is possible to see where strategic investments could reduce the cost burden of disease for society as a whole, in terms of care delivery and financing. By treating these required therapies as a cause to be managed rather than as a long-term investment in the health of society, governments in Europe and elsewhere have enacted policies that undermine health.

The healthcare system in the US has also had its share of problems. The system is under extreme pressure, but it is not coincidence that only five of the world's current 20 top-selling medicines were discovered by companies based in Europe and Japan, where governments maintain tight controls on spending for new medicines and thereby inadvertently undermine an active science base in a free market. The other 15 medicines of the top 20 have emerged from America. There are indeed many problems in the US, but the market there is more receptive and the climate is better for biomedical investment and research. Consequently, Europe's economic competitiveness may well be determined by its ability to innovate. The EU research commissioner recently stated, in January 2005, that "… overall progress in increasing Europe's research investment is far too slow as member state investment targets are too often put aside, policy measures do not go far enough or they are not pursued".

3.5.1
Aging and Longevity

The situation in Europe is especially relevant with regard to medical challenges and costs related to aging and longevity. In the year 2050, it is predicted that four of every 10 people in Italy and Japan will be aged over 60 years, and that some European countries will have more citizens aged over 80 than under 20 (Figure 3.4). The convergence of rising longevity and declining fertility carries enormous economic and social implications, including more medicines, more hospitalizations, and an increased demand for healthcare. These pressures will not be sustainable in future decades unless the challenges and opportunities of aging are met in a dramatically different way. The key to this is the formation of new partnerships, and an important part of a new approach which links all components of healthcare. One such example is a partnership between the International Longevity Center and Pfizer. This non-profit organization conducts research and advises scholars and thought leaders on how to maintain well-being and productivity throughout life and into old

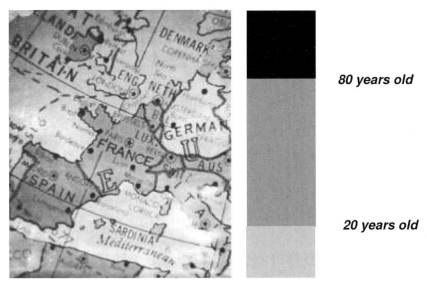

80 years old

20 years old

Figure 3.4 Predicted age distribution in European countries by the year 2050. Note that there are more citizens aged ≥ 80 years than aged ≤ 20 years.

age. At Pfizer, the phenomenon of aging is viewed as a major opportunity to rethink how societies view the entire range of health issues. In this regard, policy makers in the EU and national governments are being engaged to emphasize the link between health and overall economic benefits to society, by working together to build productive alliances and partnerships. By proving that economic benefit comes with early diagnosis and early treatment, the aim is to achieve shared goals. Currently, this spirit of partnership is also producing excellent results in the so-called neglected diseases of the developing world.

3.6
The "10–90 Gap"

It has been claimed that only 10% of global health research is devoted to conditions which account for 90% of the global disease burden: this is referred to as the "10–90 gap". I would like to be a little provocative and argue that there is no 10–90 gap, and that in poor countries the problem is one of inadequate access and infrastructure to deliver currently available drugs. The fact is that private companies are responsible for discovering, developing and producing virtually all medicines currently on the World Health Organization essential drug list, and industry continues to discover and develop many effective and safe medicines that address significant health problems in the developing world. These range from infectious diseases such as HIV/AIDS, tuberculosis, malaria, or tropical parasitic illnesses to non-communicable conditions that are now emerging as a serious double burden for the populations of the developing world. Moreover, as the lifespan of the population in developing countries continues to increase, the non-communicable diseases of the developed world will become major scourges of disease within the developing world. One recent example was the discovery and development of novel products to combat malaria. At Pfizer, one such drug is currently undergoing Phase III clinical trials, whilst several other companies have established research centers focusing on tuberculosis and other infectious diseases. It is important that this approach is continued and conducted in partnership with governments, with nongovernmental organizations (NGOs), and in collaboration with other companies.

3.7
Incentives to Stimulate Innovation

There is a need to create incentives to speed the discovery and development of medicines, and to address existing gaps in developing countries (Figure 3.5).

► **Public-Private Partnerships**

► **Advanced Purchasing Commitments**

► **A Global Fund for Tropical Diseases**

► **Tropical Diseases Drug Act**

Figure 3.5 Incentives to stimulate innovation.
(Source: Addressing the Health Needs of the Poorest Populations:
Lessons Learned and Remaining Challenges in the Fight Against Neglected Disease.
International Federation of Pharmaceutical Manufacturers Association, October 2004).

Examples of these incentives include public–private partnerships that attract funding to finance applied research in areas that are currently unattended. Typical examples include the Global Alliance for TB, and the Medicines for Malaria venture, whilst others are developing a growing portfolio of agents against tropical diseases.

A second incentive is that of advanced purchasing commitments, which represents an attempt to correct an absence of sustainable funding mechanisms and an effective market in countries that are too poor to purchase any innovative drugs. These advanced purchasing commitments create the promise that new medicines will be purchased and that R&D efforts in tropical diseases will be worthwhile. In other words, a set price and volume provide the necessary reward for the R&D effort in lieu of a real market.

A global fund for tropical diseases should also be considered to ensure that purchasing funds are made available to countries and communities that need medicines. Moreover, these funds should be made available at very low cost. An example of this is the Global Fund against AIDS, tuberculosis, and malaria, a demand-driven model which comprises a consortium of public and private sector NGOs that develops and submits grant applications.

Finally, a Tropical Diseases Drug Act, similar to the Orphan Drug Act, could be introduced, although key to the success of this would be the introduction of partnerships. The biomedical industry with their experience in the "real world" could shape the role of these new partnerships. Business should provide the lion's share of funding, and drive the emergence of a more efficient, more effective paradigm for biomedical research – a paradigm that should have a huge impact on the three A's. By sharing resources and infrastructures, these partnerships would also limit the risk to all partners. Each new project could be responsive to a wider variety of needs and a richer array of insights and more institutions. As a result, organizations would have ownership over some part of the process, but more people would have a stake in its success.

3.8
Treatment or Prevention?

Many people in the developed world will reach the age of 85 years, and by that time, based on current statistics, half of them will have Alzheimer's disease. At present, therapies exist to treat the symptoms, but not to alter the progression of the disease; an example of this is Aricept. The question is that, given the choice, should investment now be made in very difficult studies, with huge numbers of failures, in order to prevent progression of this disease? The alternative is to start building institutions that will house the millions of people aged 85 and who will have dementia. Clearly, the same choice applies to a whole range of diseases and, whilst there are no guarantees, it is likely that the incredible advances in biomedical research will provide effective treatments that will not only dramatically improve medical outcomes but also reduce the cost to society.

So, in response to the question posed in the title of this module, of whether investments in new therapies are paying off, the reply is undeniably, yes.

General Bibliography and Suggested Reading

1 Cancer Genome Anatomy Project website. http://cgap.nci.nih.gov.

2 Cancerquest website. http://www.cancerquest.org.

3 B. A. CHABNER, T. G. ROBERTS, Chemotherapy and the war on cancer (**2005**), *Nat. Rev. Cancer* 5, 65–72.

4 R. DUNCAN, The dawning era of polymer therapeutics (**2003**), *Nature Rev. Drug Disc.* 2, 347–360.

5 M. FERRARI, Cancer nanotechnology: opportunities and challenges (**2005**), *Nature Rev. Cancer* 5, 161–171.

6 D. G. FRYBACK, B. M. CRAIG, Mesuring economic outcomes of cancer (**2004**), *J. Nat. Cancer Inst. Monogr.* 33, 134–141.

7 A. KAMB, Opinion: What's wrong with our cancer models? (**2005**), *Nat. Rev. Drug Discov.* 4, 161–165.

8 K. LINDPAINTNER, Science and society: The impact of pharmacogenetics and pharmacogenomics on drug discovery (**2002**), *Nat. Rev. Drug Discov.* 1, 463–469.

9 meniscus.com website, E. ROWINSKY, *Targeting angiogenesis.* http://www.meniscus.com/horizons/5-2.pdf.

10 National Cancer Institute Alliance for Nanotechnology in Cancer website. http://nano.cancer.gov.

11 J. S. ROSS, D. P. SCHENKEIN, R. PIETRUSKO, M. ROLFE, G. P. LINETTE, J. STEC, N. E. STAGLIANO, G. S. GINSBURG, W. F. SYMMANS, L. PUSZTAI, G. N. HORTOBAGYI, Targeted therapies for cancer (**2004**), *Am. J. Clin. Pathol.* 122, 598–609.

12 B. W. STEWART, P. KLEIHUES, *World Cancer Report* (**2003**), WHO (World Health Organisation). IARC (International Agency for research on Cancer).

13 R. L. STRAUSBERG, A. J. G. SIMPSON, L. J. OLD, G. J. RIGGINS, Oncogenomics and the development of new cancer therapies (**2004**), *Nature* 429, 469–474.

14 G. M. WHITESIDES, The "right" size in nanotechnology (**2003**), *Nature Biotech.* 21, 1161–1165.

15 W. R. DOWDLE, D. R. HOPKINS (Eds.), *The Eradication of Infectious Diseases* (**1998**), John Wiley & Sons.

Health for All?: Analyses and Recommendations
Edited by The World Life Sciences Forum – BioVision
Copyright © 2005 Wiley-VCH Verlag GmbH & Co. KGaA, Weinheim
ISBN: 3-527-31489-X

16 A. ACHARYA, J. L. DIAZ-ORTEGA, G. TAMBINI, C. DE QUADROS, I. ARITA, Cost-effectiveness of measles elimination in Latin America and the Caribbean: a propsective analysis (**2002**), *Vaccine* 20, 3332–3341.

17 H. CARABIN, W. J. EDMUNDS, M. GYLDMARK, P. BEUTELS, D. LÉVY-BRUHL, H. SALO, U. K. GRIFFITHS, The cost of measles in industrialised countries (**2003**), *Vaccine* 21, 4167–4177.

18 CDC (Center for Diseases Control), *Epidemiology and Prevention of Vaccine-Preventable Diseases*. The Pink Book, 2nd printing, 8th edition (**2005**).

19 CDC (Center for Diseases Control) website, W. R. DOWDLE, *The Principles of Disease Elimination and Eradication*. http://www.cdc.gov/mmwr/preview/mmwrhtml/su48a7.htm.

20 I. CHABOT, M. M. GOETGHEBEUR, J.-P. GRÉGOIRE, The societal value of universal childhood vaccination (**2004**), *Vaccine* 22, 1992–2005.

21 GAVI (Global Alliance for Vaccines & Immunization) website. http://www.vaccinealliance.org.

22 C.-J. LEE, L. H. LEE, C.-H. LU, *Development and evaluation of drugs from laboratory through licensure to market*, 2nd edition (**2003**), CRC Press.

23 J. B. MCCORMICK, S. FISHER-HOCH, *The virus hunters: Dispatches from the Front Line* (**1997**), Bloomsbury.

24 C. A. NEEDHAM, R. CANNING, *Global disease eradication: the race for the last child* (**2003**), ASM Press.

25 Pan American Health Organization (PAHO) website, *A culture of prevention: a model for control of vaccine preventable diseases. XVI meeting of the technical advisory group on vaccine-preventable diseases*. http://www.paho.org/English/AD/FCH/IM/ TAG16_FinalReport_2004.pdf.

26 E. PEGURRI, J. A. FOX-RUSHBY, W. DAMIAN, The effects and costs of expanding the coverage of immunisation services in developing countries: a systematic literature review (**2005**), *Vaccine* 23, 1624–1635.

27 Polio Eradication Initiative website, *Global Polio Eradication Initiative. Financial resources requirements 2005–2008*. http://www.polioeradication.org/content/general/ FRR2005-2008FinalEnglish.pdf.

28 WHO (World Health Organization) website, Department of Vaccines and Biologicals, *State of the world's vaccines and immunization*. http://www.who.int/vaccines-documents/DocsPDF04/wwwSOWV_E.pdf.

29 WHO (World Health Organisation), *Report. Global Tuberculosis control: surveillance, planning, financing* (**2004**).

30 World Health Organization website, *Communicable disease Surveillance and Response*. http://www.who.int/csr/en.

31 A. C. TWADDLE (Ed.), *Health care reform around the world* (**2002**), Auburn House.

32 Global Health Council, *Technical report. Reducing Malaria's Burden: Evidence of Effectiveness for Decision Makers.* http://www.globalhealth.org/view_top.php3?id=384.

33 IFPMA (International Federation of Pharmaceutical Manufacturers Associations) website, *Research and development for neglected diseases. Lessons learned and remaining challenges* (**2004**). http://www.efpia.org/2_indust/IFPMAbrochnegdis2004.pdf.

34 International Policy network website, J. MORRIS, P. STEVENS, A. VAN GELDER, *Incentivising research & development for the diseases of poverty.* http://www.policynetwork.net/main/content.php?content_id=24.

35 P. H. SULLIVAN, *Profiting from Intellectual Capital: Extracting Value from Innovation* (**2005**), Wiley (Intellectual property series).

36 C. WHEELER, S. BERKLEY, Initial lessons from public – private partnerships in drug and vaccine development (**2001**), *Bull. WHO* 79, 728–734.

37 World Trade Law website, *Declaration on the TRIPS Agreement and Public Health.* http://www.worldtradelaw.net/doha/dohatexts.htm.

Module II
Developing, Manufacturing and Using Vaccines:
Which one is most Critical?

Health for All?: Analyses and Recommendations
Edited by The World Life Sciences Forum – BioVision
Copyright © 2005 Wiley-VCH Verlag GmbH & Co. KGaA, Weinheim
ISBN: 3-527-31489-X

Introduction

Dominique Lecourt

Growth of biotechnologies for the production of new vaccines has occurred through several stages. From traditional vaccines (poliomyelitis, tetanus, tuberculosis, measles ...), we have advanced to more sophisticated ones such as that against hepatitis B or meningitis. Today, a whole pallet of new prophylactic and therapeutic vaccines are in various stages of development, including vaccines against malaria and cervical cancer caused by the papilloma virus. The genetic constitution of the viruses appears extremely promising. How can we make these vaccines, both old and new, accessible to the greatest number of people? Although initiatives such as the Global Alliance for Vaccines and Immunization (GAVI) are leading the way, their progress also points to the obstacles. Foremost among these are economic and political barriers due to the lack of medical infrastructure in many countries where millions of children could be saved by vaccines. The lack of financial resources is often as important as the lack of political will. Perhaps more alarming is the distrust of vaccination by the general public. It is urgent that politicians in developing countries regard vaccination as well as education as a priority. In order to reduce the inequalities between North and South, shouldn't one consider producing the necessary vaccines in the same countries where their lack is felt so acutely? The fact is that mortality rates resulting from pathogens to which vaccines are targeted are much higher in developing countries than in developed countries.

Health for All?: Analyses and Recommendations
Edited by The World Life Sciences Forum – BioVision
Copyright © 2005 Wiley-VCH Verlag GmbH & Co. KGaA, Weinheim
ISBN: 3-527-31489-X

Author Biography

Sir Gustav Nossal

Professor Emeritus, Department of Pathology, University of Melbourne

Gustav Nossal was born in Bad Ischl, Austria, in 1931, and came to Australia with his family in 1939. He studied Medicine at The University of Sydney and, after residency at Royal Prince Alfred Hospital, took his PhD at The Walter and Eliza Hall Institute of Medical Research in Melbourne. Apart from two years as Assistant Professor of Genetics at Stanford University, one year at the Pasteur Institute in Paris, and one year as a Special Consultant to the World Health Organization, all Nossal's research career has been at the Hall Institute, of which he served as Director (1965–1996). Nossal was also Professor of Medical Biology at The University of Melbourne.

Nossal's research is in fundamental immunology, and he has written five books and 530 scientific articles in this and related fields. He has been President (1986–1989) of the 30,000-member world body of immunology, the International Union of Immunological Societies; President of the Australian Academy of Science (1994–1998); a member of the Prime Minister's Science, Engineering and Innovation Council (1989–1998); and Chairman of the Victorian Health Promotion Foundation (1987–1996). He has been Chairman of the committee overseeing the World Health Organization's Vaccines and Biologicals Program (1993–2002) and Chairman of the Strategic Advisory Council of the Bill and Melinda Gates Children's Vaccine Program (1998–2003).

4
Developing Affordable Vaccines

Gustav Nossal

4.1
Introduction

Before highlighting the rich promise of biotechnology in the development of new and improved – and, above all, affordable – vaccines, the question should perhaps first be asked, what is a vaccine? The essential prerequisites for a vaccine are straightforward (Figure 4.1). From a technical standpoint there must be vaccine molecules, termed antigens, and these are frequently present on the surface of a pathogen. There must also be a "danger" signal to alert the immune system that something is afoot. Then, there must be three healthy sets of cells collaborating with one another in order to set the immunological cascade in motion. The scavenger cells (technically known as dendritic cells) must be present to take up the antigen, become activated, and in turn activate a cell known as a T lymphocyte. The latter cell has a dual function: first, it attacks infected cells; and second, it helps the antibody-forming cells, the B lymphocytes, rapidly to synthesize specific antibodies.

Essential Pre-requisites for a Vaccine

▶ Vaccine molecules (antigens), frequently from the surface of a pathogen.

▶ A "danger" signal to alert the immune system.

▶ Three healthy sets of cells collaborating with one another:
  - dendritic or scavenger cells to take up antigen;
  - T lymphocytes to attack infected cells;
  - B lymphocytes to make specific antibodies.

Figure 4.1 Essential prerequisites for a vaccine.

Health for All?: Analyses and Recommendations
Edited by The World Life Sciences Forum – BioVision
Copyright © 2005 Wiley-VCH Verlag GmbH & Co. KGaA, Weinheim
ISBN: 3-527-31489-X

4.2
The "Danger" Signals

With regard to the nature of these danger signals, it appears that an ancient, evolutionary system of receptors on scavenger cells (among others) was designed to recognize pathogen-associated molecular patterns. This is probably the prime example of a danger signal. In addition, proteins can be released from dying cells following viral infection or tissue damage as a result of inflammation, and this generates the immune response. In the laboratory, the danger signals are artificially created by preparing chemical mixtures, termed adjuvants, that may contain several such activators. Harmless attenuated viruses may themselves signal danger and may be genetically engineered to produce several antigens.

4.3
Bioengineered Vaccines

Several well-known examples of bioengineered vaccines have been prepared through modern biotechnology (Figure 4.2). One such example is the hepatitis B surface antigen (HBsAg), which is currently prepared in yeast cells. These molecules self-aggregate and self-assemble into virus-like particles that are attracted towards scavenger cells, thereby achieving major vaccination success against a very serious pathogen.

In a more complex scenario, involving GlaxoSmithKline (GSK) and the US Department of Defense, the malarial antigen from the surface of the parasite as it leaves the mosquito (the so-called circumsporozoite protein) is fused to the HBsAg. In this way, a virus-like particle is formed that can be mixed with strong adjuvants such as RTS or S-ASO2. After almost two decades of

---

Interesting Examples of Bioengineered Vaccines

▶ Hepatitis B surface antigen (HBsAG) made by yeast cells.

▶ The malarial antigen from the surface of the parasite as it comes out of the mosquito (CSP) fused to HBsAg to form virus-like particles, mixed with strong adjuvant chemicals.

▶ The polysaccharide-protein conjugate vaccines against meningitis (such as Hib and Mening C).

Figure 4.2 Examples of bioengineered vaccines.

problematic research, good partial protection is now available in African children, and the proof of principle that the malaria community so ardently desired has clearly been provided.

The polysaccharide–protein conjugate vaccines represent an intelligent option as they utilize T-cell/B-cell collaboration. In the initial step, the B cells recognize the contact-specific polysaccharide, which is then conjugated to a protein. This appears to "annoy" the helper T cells, and results in a much better response than if the polysaccharide alone were present. The first such vaccine to be introduced in practice was *Haemophilus influenza* B, and this proved highly successful against meningitis. A vaccine against *Streptococcus pneumoniae* has only recently been developed. The results of a huge, four-year study involving 17 000 young children in The Gambia, West Africa were published in *The Lancet* (25th March, 2005). The nine-valent vaccine was seen to be 77% effective in preventing pneumococcus-derived pneumonia. Moreover, it reduced the incidence of pneumonia from all causes by 37%, and reduced all-cause mortality by 16%. In other words, a vaccine directed against acute respiratory disease is saving children's' lives incontrovertibly, and with very high statistical significance.

4.4
Biotechnological "Tricks" in Vaccine Production

Most widely used, currently available vaccines have been developed on the basis of good antibody production. After all, this is the target of the researcher in preclinical studies, and is what happens in experimental animals. Although antibodies are produced by B cells, some infections must be combated by T cells attacking the infected cells and secreting protective molecules associated with the inflammatory process. Two key examples of this are tuberculosis and HIV/AIDS. One problem with this is that the sustained activation of cytotoxic T cells has been difficult to achieve, especially in subhuman primates and humans. Thus, a number of biotechnological "tricks" have been developed (but not yet perfected) to promote T-cell activation. The first is to use viruses that have been genetically engineered to ferry pathogen antigens into the body; this is the "Trojan horse" concept (Figure 4.3).

The original virus used widely in experimental animals is vaccinia (the smallpox virus) and its variants, which are double-stranded DNA viruses. Adenoviruses (also double-stranded DNA viruses) have been widely used, as have variants of the AIDS virus itself, which are retroviruses and of course single-stranded RNA viruses. In addition to using the Trojan horse to carry in the antigenic pathogen, it is also possible to co-express with the antigen certain stimulatory molecules, termed cytokines. This makes the immune response stronger, and even naked DNA encoding antigens can be used. In fact, some

Some Biotechnology Tricks Promoting T Cell Activation

▶ Use of viruses genetically engineered to ferry pathogen antigens into the body (the Trojan horse concept).

- Vaccinia and variants (DS-DNA)
- Adenoviruses (DS-DNA)
- Retroviruses (SS-RNA)
- Co-expression of cytokines

▶ Use of "naked" DNA encoding antigens; some formulations include a scavenger cell targetting strategy.

▶ "Prime-boost" strategies:

- DNA prime – protein boost
- DNA prime – viral boost

Figure 4.3 Biotechnology tricks promoting T-cell activation.

recent formulations of DNA vaccines have included a scavenger cell targeting strategy, where the antigen of molecules for which the scavenger cells have receptors is attached to the DNA.

The problem is that this does not achieve a strong and consistently long-lasting cytotoxic T-cell response. Consequently, it was decided to focus the response on the T cell by using a "prime-boost" strategy – that is, DNA priming with either a protein boost or viral boost. This approach, although attractive, is currently only at the experimental animal stage or in very early Phase I clinical trials, and is far from complete.

4.5
Needle-free Immunization

4.5.1
Mucosal Immunization

The concept of needle-free immunization has been under consideration for some time, with mucosal immunization being a major contender. In laboratory animals, vaccines can function effectively when introduced via mucosal surfaces, including oral, intra-nasal, respiratory, and rectal surfaces. The problem in this situation is that as well as the antigen of interest, a strong mucosal adjuvant is also required; typical examples include the B subunit of cholera toxin, mutants of the *Escherichia coli* heat-labile toxin or camptothecin (CPT) conjugate oligonucleotides. If such a mucosal adjuvant is available, it can be covalently coupled to the antigen, thereby reducing the quantity of antigen required, which is a very practical point. Unfortunately, although the mucosal immunization strategy shows great promise, very few supportive clinical data are currently available.

4.5.2
Transdermal Immunization

Transdermal immunization represents an alternative form of needle-free immunization. The dendritic cells lie quite superficially in the skin, and model antigens are able to penetrate the skin, especially if it has been heavily pre-moistened. Again, an adjuvant is required for immunogenicity, and the development of such materials has led to considerable interest among smaller biotechnology companies. The main interest is in formulations which allow better penetration of proteins into the skin. As an alternative, some companies are producing very fine, almost brush-like, needle-like arrays in patches. These can be glued onto the skin, avoiding needle pricks, but are still at the experimental stage.

4.6
Genome Mining

Genome mining represents a new concept of antigen development. Today, the entire genome sequence of many pathogens, viruses, bacteria and parasites is available. These data can be analyzed using bioinformatic software to identify genes which code for surface-exposed proteins, virulence factors, molecules involved in cell invasion, or other vaccine candidate proteins. Moreover, by using recombinant DNA technology, the relevant proteins can be easily synthesized and individually tested in mice. As an example of this reverse vaccinology, Chiron has tested 350 proteins of *Neisseria meningitidis* group B. Although as yet there is no freely available vaccine against meningitidis B, 28 novel antigens have been identified which elicited bactericidal activity in mice. Currently, the Chiron group is patiently working through these antigens, hoping to develop a vaccine "cocktail" that is effective in humans.

4.7
Vaccine Affordability

Two examples spring to mind of how the world is addressing the problem of vaccine affordability. The first of these relates to rotavirus, which was discovered by Ruth Bishop and Ian Holmes in Australia some 30 years ago, and is the most powerful viral cause of diarrhea among infants. In fact, rotavirus kills some 600 000 infants each year. Recently, in January 2005, GSK launched its oral live attenuated single strain G1 Rotarix vaccine in Mexico. Although not hot news, this was the first occasion that a multinational corporation had gone to a developing country where the problem lay, had used that country's regulator to obtain the first registration, and negotiated with that country to introduce the vaccine into the population. Jean Stephenne will discuss this vaccine in more detail in Chapter 6. GSK has announced that there will be three levels of tiered pricing for this vaccine: a high price if it is eventually introduced into wealthy countries; a medium price for countries in transition (e.g., Mexico); and a low price for those countries that are the poorest of the poor.

In this respect, GSK is not alone, with Merck having studied more than 70 000 infants in 11 countries after administration of their oral, three-dose human-bovine reassortant vaccine containing five strains, the so-called RotaTeq. The vaccine was 74% effective against any diarrhea, impressively (98%) effective against severe diarrhea, and was also well tolerated.

There is an emerging research capability in the developing world itself, with India and Indonesia both having interesting rotavirus vaccine candidates. However, these are not as far advanced along the research pipeline as the vaccines, which are already in use. Particular interest has been expressed in the Indonesian version (this is Ruth Bishop's original strain), although as yet the clinical efficacy has not been monitored.

4.7.1
The *Neisseria meningitidis* Serogroup-A Conjugate Vaccine

In the case of African meningitis, the *Neisseria meningitidis* serogroup-A conjugate vaccine represents a completely different answer to the affordability conundrum. It is well known that, every three to four years, vicious epidemics of meningitis sweep across the African meningitis belt from Senegal to Ethiopia, threatening 300 million people. The cause usually (but not always) is meningococcus A, and a US$ 17 million grant from the Gates Foundation has proposed, for competition, the idea of a new monovalent group A meningitis vaccine. The best tender came from the Serum Institute of India in Pune, which has guaranteed to supply the vaccine at US 40 cents per dose. There is an interesting strategy in operation here, as the polysaccharide antigen and carrier tetanus toxoid will each be provided by manufacturers from an industrialized country, whereas the conjugation technology utilized will be transferred through an elaborate contracted collaborative venture. This major experiment is the responsibility of Marc LaForce, the eminent Canadian scientist leading the WHO Gates Foundation team. If it is successful, the low labor costs and great energy of developing country manufacturers can be used in the search for new and improved vaccines.

4.8
The Next Challenge

The next major challenge to the biotechnology of vaccinology is, most likely, that of more combination treatments and fewer needle injections (Figure 4.4). There is a need to combat the fact that, in many developing countries, disposable syringes are reused and so can transmit blood-borne diseases. Consequently, the introduction of auto-disable and uniject syringes is critical. Moreover, there is a need for vaccines that do not require refrigeration, or that would function earlier in life. An example of this is the famous "gap" in measles, where maternal immunity wanes at 4–6 months, but the available vaccine cannot be administered before the child is aged 9 months. It has been argued that unless a so-called "stealth" vaccine can be produced, which will "creep" under the residual antibody but still be able to protect 4-month-old children, measles might never be eradicated from the world. In Chapter 2, Ciro de Quadros suggested that, although this situation is not necessarily true if a highly disciplined approach is followed, vaccines which functioned earlier in life would be of great value. Clearly, the "holy grail" is vaccines for the big three diseases – HIV/AIDS, tuberculosis, and malaria – but these are still some distance away.

Other Biotechnology Challenges in Vaccinology

▶ More combinations, fewer needle-pricks.

▶ Autodisposable and uniject syringes.

▶ Vaccines which do not require a cold chain.

▶ Vaccines which function earlier in life (e. g. measles).

▶ Vaccines for the big 3 – HIV/AIDS, tuberculosis, malaria.

Figure 4.4 Biotechnological challenges in vaccinology.

4.9
Conclusion

To conclude on a note of hope, it would be nice to think that 2005 – the 50th anniversary of the discovery of the Salk vaccine – will be a historic turning point in how the question of producing vaccines for the developing world is approached. I would remind you of the comment made by the Irish poet Nobel laureate Seamus Heaney, when he heard that Nelson Mandela had at long last been released from jail: "A further shore is reachable from here. Once in a lifetime justice can rise up and hope and history rhyme".

Author Biography

Jacques-François Martin

Member of the Board, GAVI, President, Global Fund for Children's Vaccines

Jacques-François Martin has spent the essence of his career in the pharmaceutical, biological and life sciences industries.

From 1970 to 1976, he was the Chief Executive Officer of *Rhône-Poulenc Pharma* in Hamburg, Germany. He then returned to France to join the *Institut Mérieux* as Vice President of sales and marketing, where he largely contributed to the international expansion of Mérieux. He was named the company's Chief Executive Officer in 1988, and successfully negotiated with the government of Canada to acquire *Connaught Laboratories*.

In 1991, J.-F. Martin set up *Parteurop S.A.*, a biotech consulting company based in Lyon, France. At Parteurop, as Chairman and CEO, he helps establish start-up companies by leveraging innovation from French and foreign institutions.

From 1996 to 1998, he was the Chief Executive Officer of the *Fondation Jean Dausset – Centre d'Etudes du Polymorphisme Humain*, a private foundation dedicated to genomics research.

From September 1996 to September 1999, he was a member of the Board of *INSERM* (Institut National de la Santé et de la Recherche Médicale, the French National Institute of Health).

From 1994 to 1997, he served as the Chairman of the *Biologicals Committee of the International Federation of Pharmaceutical Manufacturers Associations*. As such, he was a member of the Scientific Advisory Group of Experts of WHO (SAGE).

From November 1997 to June 2003 he was a member of the Board of the International AIDS Vaccine Initiative (*IAVI*).

From 2000 to January 2005, J.-F. held the position of President of *The Vaccine Fund*. He lead the Fund's efforts to provide lifesaving vaccines and other immunization program support to low-income countries. In this capacity, he was also a member of the Board of *GAVI* (Global Alliance for Vaccines and Immunization).

He is member of the Board of several life sciences companies.

J.-F. Martin holds a Master's in Business Administration from the Ecole des Hautes Etudes Commerciales.

5
Why "Affordable" Vaccines are Not Available
to the Poorest Countries

Jacques-François Martin

5.1
Introduction

Today, work is progressing on many new vaccines, the need for which is unquestioned. There is, however, a paradox. At a meeting in Hanoi which centered on diseases (notably diarrheal) of the most impoverished nations, much concern was expressed that following the tsunami on 26th December 2004, enteric diseases prevalent in some of the affected countries might be dispersed, leading to serious epidemics. This was particularly the case for typhoid fever and cholera. The question was also asked that if these diseases are endemic in the countries referred to, and good vaccines are available to prevent cholera and typhoid fever, then why are we not using them?

We live in a strange world when it comes to vaccines. We are dreaming of new vaccines that are not yet available, whilst those that are available are not used appropriately. It was this reflection which led to the idea of renovating and revamping efforts to immunize children through the Global Alliance for Vaccines and Immunization (GAVI), which was started five years ago.

When this idea was first conceived, the situation was that among each annual birth cohort of about 130 million children, 36 million had no access to immunization. As a consequence, between two and three million children would die from a disease which could have been prevented by existing vaccines. If calculated on a monthly basis, three million per year approximates, in terms of death, a tsunami every month. In total, 250 000 children were dying – some say unnecessarily, I would say scandalously – from diseases which could be relatively easily prevented.

Why has this happened? Is it too difficult to develop the required immunization strategies? That is definitely not the case. We are all aware of the fantastic efforts deployed during the 1980s, when coverage in terms of immunization was very poor, through the early 1990s when, under the extraordinary

Health for All?: Analyses and Recommendations
Edited by The World Life Sciences Forum – BioVision
Copyright © 2005 Wiley-VCH Verlag GmbH & Co. KGaA, Weinheim
ISBN: 3-527-31489-X

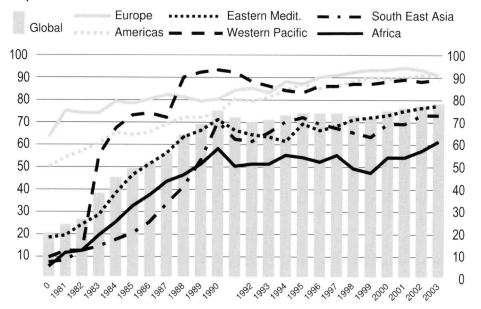

Figure 5.1 DTP3 coverage (in %), 1980–2003. (Source: WHO/UNICEF).

leadership of UNICEF and particularly, Jim Grant, substantial progress was achieved (Figure 5.1). During this period, whilst initially 10–20% of children were immunized with six basic vaccines, this figure ultimately rose to 60–80%, depending on the continent. Subsequently, and for a variety of reasons, the effort has not been sustained in a way to maintain, let alone improve these results.

In a nutshell, we have a track record of success with immunization, and we know how to do it. Still immunization is not performed appropriately. Is that because it is not cost efficient and we should have other priorities in healthcare? When comparing different health activities, in terms of lives saved per million dollars invested, immunization is among the most cost efficient (Table 5.1).

Table 5.1 Cost efficiency of immunization versus other health services for children. (Source: WHO, 2003).

Intervention	Lives saved per million US$ spent
Enhanced program of vaccination (six standard antigens)	1500–2500
Control of malaria (different measures)	1200–1500
Treatment of cancer	2–10
Preventing HIV mother-to-child transmission	2500–5000

So, there is a medical need, the know-how is available, there is a track record of success, and the return on investment is among the best, why do these problems still exist? Several reasons have been suggested, but I will focus on three main elements, namely political commitment, access to vaccines, and the financial resources available.

5.2
Political Commitment

Immunization is an activity which must be sustained and systematized, because children are born continuously. Generally, immunization is carried out at the discretion of health clinics worldwide; it is not a spectacular activity and therefore requires constant political commitment to ensure that it is conducted as it should be. In many countries, health budgets are inadequate, without even considering development levels. For example, the African Union has suggested that African countries should spend 15% of their national budget on health in general. At present, this figure is, on average, 5%, so there is still room for progress. In both less-developed and developed countries, the notion of prevention is less popular than that of treatment, although it is much cheaper and much more efficient to prevent than to treat. Likewise, immunization does not sell well to politicians, because it is much more "spectacular" to open a new hospital than to support a continuous immunization activity. Overall, it appears that there is insufficient political commitment in many countries, and this was the situation some five years ago.

With regard to international organizations, institutions such as UNICEF and WHO have many different activities to develop in many fields, and so cannot be expected to have full-level priorities for all cases. For UNICEF, immunization has always been a major program, although the level of priority has changed with time. It is important to ensure that the appropriate organization maintains the momentum for immunization.

In the case of donations, the situation has arisen where donors expressed concern about how their money would be used, and what results would be achieved. A somewhat sterile debate was also continued on the benefits of a program approach versus a horizontal approach. Clearly, each health activity must be examined in the context of the broader health sector, as no one clinic worldwide performs only immunization. It is also clear that when children are immunized, access to many other health schemes is secured. The lack of political commitment in immunization is the reason why the GAVI was started five years ago – to motivate and coordinate activities between different players at national and international levels, to advocate the cause, to mobilize resources and, generally speaking, to reposition immunization as a centerpiece in the design and assessment of international development efforts.

5.2.1
The Results to Date

In terms of results achieved to date, the delivery of vaccine doses has clearly risen (Figure 5.2). We are particularly proud of the progress of hepatitis B (Hib) vaccine, which was launched in wealthy countries about 20 years ago, but has been used appropriately in developing countries only during the past five years. Today, this vaccine has been introduced into most developing countries, and basic vaccines such as DPT (diphtheria, pertussis, tetanus) have been improved. By contrast, the results with Hib are still probably insufficient; at the end of 2003, six million more children were reached with basic vaccines, and four to five million were reached with new vaccines, which is a clear breakthrough. The WHO has estimated that 700 000 deaths have been prevented by the additional immunization of these children. So, the initial number of 36 million children who remain un-immunized has, five years later, been reduced to about 27 million. Whilst this represents substantial progress, it shows that there is still much to do and that future efforts should be maintained.

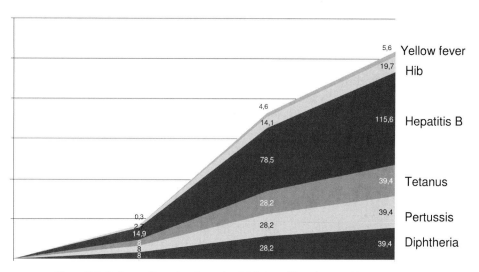

Figure 5.2 Delivery of vaccine doses by GAVI, in million doses, 2000–2003.

5.3
Access

Access to vaccination remains a major issue, and relates mainly to material points in terms of infrastructure. In particular, this applies to the cold chain (i.e., refrigeration), which is very often insufficient or old-fashioned and requires renovation. Many countries also have problems with transportation, though these apply to healthcare in general and not only to immunization. It is also important to know that progress is being made to reach inaccessible children, and not only those who are easily accessible. It appears that more effort is needed to develop outreach services when children have neither the possibility nor opportunity to visit health services. In this respect, the point must be made that in addition to its intrinsic value, immunization provides an excellent means of improving child access to many other health services. When the decision is made to fully immunize a child, it is imperative that the child attends the health service five times during the first year of life, as five injections must be given. If, in any country, 90% of newborn children are reached each year (this is the target for 2010), the indication is that there is in place the basic skeleton of a system that can be built on in order to improve access to other activities. This very proposal was the subject of a recent WHO review that dealt with the distribution of bed nets to protect against malaria at a time when children received immunization against measles.

One important part of the success of these programs is related to the issues of management and monitoring; these are very often neglected, leading in turn to substantial losses in terms of the efficiency of a country's health programs. It is also relevant at this point to mention the subject of insufficient human resources.

5.3.1
Health Worker Availability

When examining health worker density data by continent, there is seen to be wide diversity that is, perhaps, not unexpected (Figure 5.3). In wealthy countries, or where there is a tradition of global health, the average is almost 10 health workers per 1000 inhabitants. While ratios are already relatively low in Central and South America, the Middle East and Asia, the situation in sub-Saharan Africa is truly alarming, where there is less than one-tenth of the health workers present in more developed countries. On examining the potential consequences of this situation, there is seen to be a clear correlation between numbers of health workers per 1000 inhabitants and child mortality (Figure 5.4). Moreover, as long as there is no development in the health system of a country, the progress that can be made will be limited.

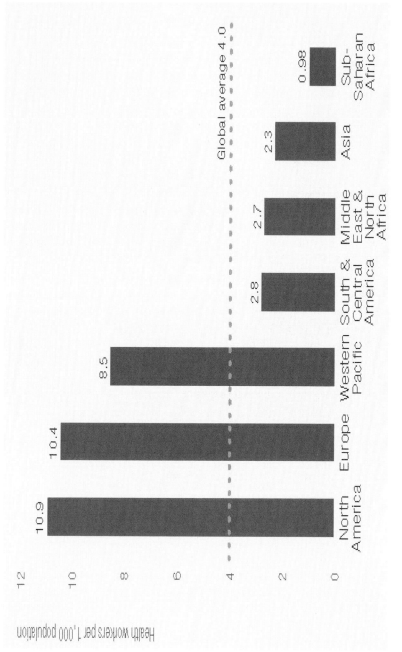

Figure 5.3 Health worker density in different parts of the world. (Source: WHO, 2004).

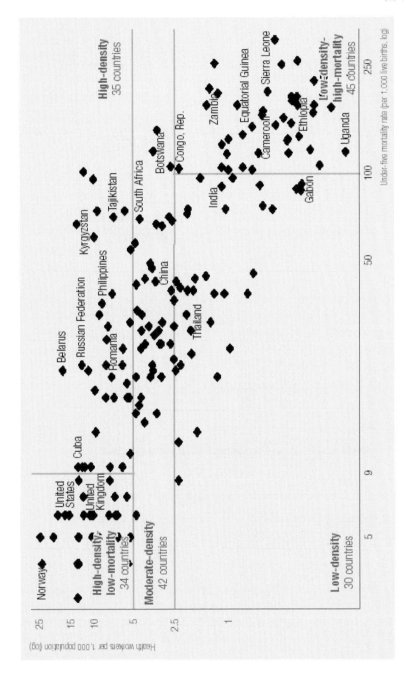

Figure 5.4 Health worker density versus child mortality in different countries. (Source: WHO, 2004 and UNICEF, 2003).

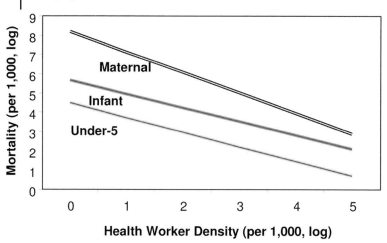

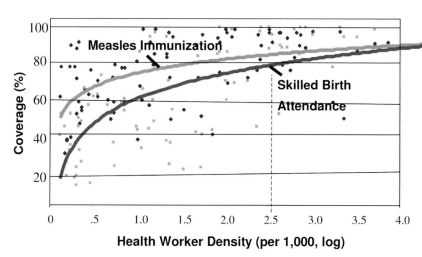

Figure 5.5 Influence of health worker density on service coverage and mortality. (Source: Anand and Baernighausen, JLI, 2004).

Infant mortality is correlated not only to the number of health workers but also to maternal mortality (Figure 5.5, upper). More precisely, immunization is also correlated to the numbers of skilled birth attendants (Figure 5.5, lower). It is difficult to improve results when numbers of health worker are insufficient, as is often the situation today. The difficulty here is that many health workers in developing countries prefer to move to wealthy countries for a better lifestyle. Indicative of this situation was an example of six African countries where, on average, 50% of health workers, if given the opportunity, would move to a wealthy country (Table 5.2).

Table 5.2 Migration intentions of health workers. (Source: WHO, 2002).

Country	Health workers who intend to migrate [%]
Cameroon	49.3
Ghana	61.6
Senegal	37.9
South Africa	58.3
Uganda	26.1
Zimbabwe	68.0

Table 5.3 Projected nursing shortfall in rich countries. (Source: U.S. Government, 2004).

Country	Projected nurse shortfall (year)	
United States	500 000	(2015)
Canada	113 000	(2011)
United Kingdom	35 000	(2008)
Australia	31 000	(2006)

The main problem is that, among wealthy countries, an insufficient number of people is trained for work in the health sector, and consequently the shortfall, in nurses alone, is considerable (Table 5.3). There is, therefore, a huge responsibility among wealthy countries to prevent such a "brain drain", with its very serious consequences for the developing world.

5.4
Financial Resources

In addition to political will, access to vaccination and an adequate infrastructure to mobilize resources, it is clear that in order to improve immunization activities there is also a need to improve the resources themselves. The figures involved are not huge. With 27 million children to be immunized each year, at a maximum average cost of US$ 30, a total of US$ 800 million is needed to save more than two million lives. The expectation is that more resources will be needed in the future as new vaccines become available, including rotavirus, pneumococcus, meningitis, and human papilloma virus vaccines. These new vaccines should be introduced during the next few years, and will be of major importance in developing countries. So, it must be ensured that, in the future, there will not be a 20-year delay between the launch of a new vaccine in wealthy and poor countries. This will be a key indicator of the efficiency of the GAVI, ensuring that in future these new vaccines can be introduced much more quickly – perhaps even at the same time – in both developing and developed countries.

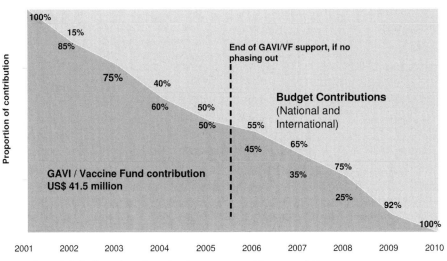

Figure 5.6 Co-financing of vaccination programs: The case of Ghana.

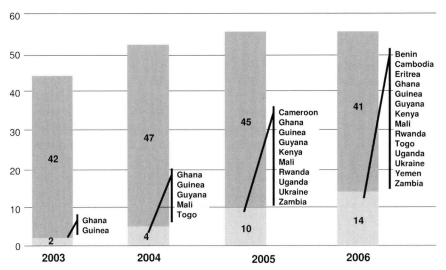

Figure 5.7 Countries co-financing vaccination programs.
Upper bar: countries without government support of the GAVI vaccination program.
Lower bar: countries which co-finance the GAVI vaccination program.
(Source: GAVI secretariat).

The global monitoring of programs will have a positive impact, and increased political commitment should lead to favorable decisions. Figure 5.7 illustrates how, in the case of Ghana, we were able to establish a process by which the country itself plans how to take over from the GAVI, and the initial input that GAVI will give to the program by planning to increase their own resources. For Ghana, the intention is, by the year 2010, to take over themselves the increased efforts on immunization induced by GAVI. The percentage of countries co-financing the vaccination effort remains very low (Figure 5.7), although there is a tendency for countries to take over a substantial part of these additional efforts. Ultimately, this should prove very beneficial for immunization activities.

5.4.1
Sources of Funding

The Vaccine Fund has a clear mission, to obtain more funds from donors, and preferably from international donors. The Fund's mission is to champion, monitor the results of, and help to sustain the efforts of the global alliance in

Table 5.4 Sources of GAVI funds.

Donor	Cumulative commitments (1999–2004) [US$ million]	Cumulative pledges (2005–2015) [US$ million]
Bill & Melinda Gates Foundation	758.5	750.0
Canada	13.0	149.5
Denmark	4.5	0.0
European Union	1.0	17.0
France	6.0	12.0
Ireland	2.0	0.0
Luxembourg	0.5	0.5
Miscellaneous private donations	4.0	0.0
Netherlands	71.0	17.0
Norway	102.0	282.0
Sweden	12.5	14.0
UK	51.0	13.5
USA	219.0	64.5
Grand total	1300	1300

order to protect children in the poorest countries. The results have been very encouraging. As of today, the Vaccine Fund has received US$ 1.3 billion from its inception five years ago, and additional commitments of the same amount of US$ 1.3 billion (of which the Gates Foundation is contributing about one-half). Whether it is right for a private foundation to represent one-half of the global effort is subject to debate, but we have commitments of another US$ 1.3 billion for the next few years, and this is important in order to have appropriate visibility (Table 5.4). Some countries are incredibly generous, notably Norway, which has only five million inhabitants. If all countries were to provide the same amount of money per capita as Norway, the situation would be excellent. However, if the current programs are to be carried out with this level of resources, the funding must be improved very substantially. This is particularly the case when financing the newer, technologically complicated vaccines that are inherently more expensive. At present, this financial situation is problematic, and remains a challenge for the GAVI.

5.4.2
The International Finance Facility

The Millennium Development Goals, which were defined by the United Nations in 2000, include as their fourth item the goal of reducing child mortality by two-thirds by the year 2015. To prevent child death through immunization is an integral part of the potential success, but today, the resources flowing to these developing countries in order to make a greater difference remain insufficient. There is a clear need to raise increasing amounts of money at a time when national budgets in many wealthy countries are very limited. The idea of an International Finance Facility (IFF) (Figure 5.8) was developed by the British Chancellor, Gordon Brown. The idea, which in principle is relatively simple, is to create a new financial instrument (the IFF) to which countries would make pledges over long periods of time, perhaps 10, 15 or 20 years. Based on these commitments from donors, the IFF would then offer the prospective of being reimbursed by pledges of future governments. This would lead to a situation in which current effort could be substantially improved whilst still having access to additional resources. These resources would be added to over time, and particularly with regard to the question of access, it would represent a very important mechanism. At the beginning of 2005, the richest governments announced that they would commit US$ 1.8 billion over 10 years through the IFF, as long as other players matched their contribution. The French government expressed support in principle, proposing that they would secure financing of the additional commitment through an international tax. The IFF concept has raised several questions, and although it is not yet in place, much hope has been placed in its institution.

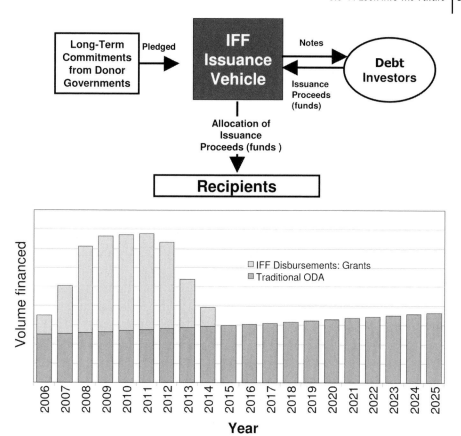

Figure 5.8 A new initiative: The International Finance Facility (IFF).

5.5
A Look into the Future

In closing, brief mention should be made of recent changes and suggested future changes to improve and consolidate these initial successes. Among three points to be raised, the first is that of accountability. The time has gone when financial means could be obtained for developing countries without organizing the system in a way that would be more accountable. Thus, it is important to measure, to check, and to report. Country leadership is also very important, as without leadership nothing can happen. The task is carried out, but in the context of a coordinated approach, much emphasis must be placed within the GAVI to review and check all applications and annual reports. There is also a clear need to ensure that donors are comfortable about how efficiently their money is being used.

The second concept, sustainability, is important because immunization is a long-term undertaking. The improvement of infrastructure is also a long-term task, and countries may encounter difficulties while depending on financial help that, in turn, depends on an annual budget. The Minister of Health in any developing country will generally not know in December how much money will be provided by international health organizations over the next year. Consequently, it is very difficult to propose activities not only for the next year but also over many years. That is why GAVI has introduced the five-year concept.

Finally, industry also requires visibility in order to justify investments in R&D and production capacities, all of which need time to be recouped. Thus, it is necessary to provide the corresponding visibility in terms of financial resources available to buy the product.

Accountability, sustainability and innovation form the basis for the GAVI. The GAVI, as a global alliance, brings together all players at national and international levels, NGOs, foundations, and scientific bodies in order to optimize the global efficiency of all concerned, and in this respect industry is an important player. The GAVI involves industry to a great degree, whereas the Global Fund to fight AIDS, Tuberculosis and Malaria prefers to take another route. Having industry on board is an important factor since, in the way that situations are managed and accountability is built up, much emphasis is placed on the notion of a contractual approach with countries based on multi-year plans. Consequently, the decisions are made by countries based on evidence, and contributions will depend on the results achieved by those countries. Last, but not least, the question of advocacy and resource mobilization should be shared by the global community.

Whilst advocacy and resource mobilization is the mission of the Vaccine Fund, we are all part of the global process to ensure that resources are made available. Clearly, progress has been made, but there is still a long way to go. The opportunity exists to provide access to basic protection for almost all children worldwide. During the past four years, Nelson Mandela has served as the chairman of the Vaccine Fund's board, and remains convinced that immunization is the basic right of every child. Moreover, he is convinced that because we have the tools to protect these children, because we know how to do it, and because the necessary resources are relatively limited, we should all – individually and collectively – bear part of the global responsibility to ensure that the task is completed.

Author Biography

Jean Stephenne

President and General Manager, GlaxoSmithKline Biologicals

Jean Stephenne has overseen GlaxoSmithKline Biologicals since 1991, serving as Vice President and General Manager, then Senior Vice President and General Manager, until his appointment as President and General Manager in 1998. Prior to this, he was Vice President of Human Vaccines Research and Development and Production from 1988 to 1991. Jean Stephenne joined the company in 1974 as head of bacterial and viral vaccines production, becoming Vaccine Production Director in 1980. He served in a variety of capacities as Vaccine Plant Director and R&D Director from 1981 to 1991. He has a degree as engineer in Chemistry and Bioindustries from the University of Gembloux (Belgium) in 1972 and obtained a degree in Management from the University of Louvain, Belgium. He has served as President of the Union of French Speaking Companies (1998–2000), is a member of the European Association of Vaccines Producers, and a member of the Management Committee of the Belgian Companies Federation. He is also a member of the Board of several Belgian companies.

6

Meeting the Challenges of Manufacture and Delivery of Affordable Vaccines

Jean Stephenne

6.1
Introduction

In the preceding chapters, much has been said of the progress being made in the development and affordability of vaccines, and how people are beginning to recognize the value of vaccines in society. Here, the subjects to be discussed are the values of vaccines, their manufacture, and R&D developments in this field.

The lack of progress in vaccination schemes that we are experiencing is due largely to an ongoing discussion of the value of vaccines, rather than to their being recognized as a "right to life", in both developed and developing countries (Figure 6.1).

Today, nobody recognizes that value, and that perception must be changed. Indeed, it is the responsibility of society to recognize the value of vaccines, and this must be based on the contributions that vaccines make to quality of life in society. Currently, 26 diseases are vaccine-preventable, and it has been

Industrialized Countries **Developing Countries**

900 million **5.0 billion**

Figure 6.1 The global vaccine market.

Health for All?: Analyses and Recommendations
Edited by The World Life Sciences Forum – BioVision
Copyright © 2005 Wiley-VCH Verlag GmbH & Co. KGaA, Weinheim
ISBN: 3-527-31489-X

on two occasions that "herd immunity" has been acquired, in the case of meningitis C and streptococcus pneumoniae. The progress is hampered, however, by what might be considered individualism in society, and why some people have the right to refuse vaccination. It is understandable that some people may refuse vaccination, but in doing so they are putting at risk others within that society, and this aspect must be recognized. In the United States there is a clear liability to vaccination, and debate will continue with regard to those who refuse vaccination and their liability towards society. Last year, GlaxoSmithKline (GSK) – and probably many other vaccine producers – spent many millions of dollars defending themselves against possible litigations in the United States. In fact, this is the main reason why the vaccine industry has largely diminished since the 1970s. The question is, can the debate be restarted in the coming years? There is indeed a risk, particularly in the United States, that debate will return, and this point must be addressed if there is a desire to defend the value of vaccines. Within the past 20 years, GSK has taken the view that the global market is not restricted to the children of Europe and United States, but includes all the children of the world. Access to vaccine must be made available to all in the future.

6.1.1
Market Challenges

The modern vaccine market faces many challenges, because when man travels through society, infectious diseases travel with him. This important point must be remembered, because only 50 years ago, if an influenza pandemic occurred, the procedure was to place infected people into quarantine. Fortunately, that situation no longer applies, but if infectious diseases are returning, then preventive measures must be considered, and vaccination surely offers the best protection. Consequently, R&D investment in vaccines will be required to supply the needs of the developed and developing world. It is essential that manufacturing follows demand, and that there are immediately available stocks of vaccine worldwide. It is for this reason that GSK proposed the launch of its rotavirus vaccine first in Mexico, in an attempt to persuade the Mexican government to accept their responsibilities and not always to complain about the vaccine industry, that there is insufficient investment in vaccine production. In industry, it is essential that reasonable margins are incorporated: where once the subject was dual pricing, it is now three-tier pricing, and this is the key to continued investment in vaccine production.

6.2
Vaccine Research and Development

Although many consider vaccines to be cheap, this is simply not true. Today, the total cost of developing a modern vaccine, and conducting the appropriate testing, is rapidly approaching US$ 1 billion (Figure 6.2). For example, the anti-malaria vaccine which began development in 1982, and if launched in 2009–2010, will have cost more than US$ 1 billion to bring to market. This level of cost requires the co-operation of others, in this case the Walter Reed Hospital and many other scientific institutions, otherwise the whole procedure cannot be conceived. The illustration in Figure 6.2 provides three messages. First, the preparation of a DNA vaccine or live vector constitutes only a very small portion of the process in relation to the overall project. Second, it is relatively easy to induce an immune response, but to obtain an effective immune response remains a challenge. Immunity is often a lot more complex than people assume, and this is convincingly demonstrated by the examples given by Sir Gustav Nossal in Chapter 4. Third, it is important not to repeat the mistakes with the HIV vaccine, when, due to public pressure, there was a rush to conduct large clinical studies with conjugated vaccines that ultimately failed. It follows that correct science must be instituted before clinical trials are considered.

In the past, the vaccine industry has suffered attacks from both the scientific community and civilian society on the basis that it was not conducting

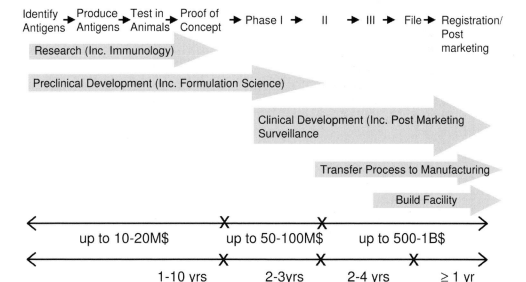

Figure 6.2 The Research & Development pipeline for new vaccines.

Pre-clinical	Phase I	Phase II	Phase III / Filed
RSV	Flu improved	EBV	N.Meningitis
CMV	HIV	Malaria	combinations
Men B (paed)	S.Pneumoniae	Hepatitis E	Rotarix™(Rotavirus)
Chlamydia	(elderly)	Dengue	Streptorix ™(Strepto)
Staph Aureus	TB		Cervarix™ (HPV)
SARS	Zoster		Priorix tetra ™(MMRV)
Cancer	Flu pandemic		Simplirix™(Herpes)
Allergy			
	Prostate Cancer	Staph. Antibodies	Boostrix™ IPV (dtpa IPV)
	Breast Cancer	(Mab)	Fendrix™ (HepB
		Lung Cancer	hemodialisys)
		Melanoma	
Total = 8	Total = 8	Total = 7	Total =8

Figure 6.3 Vaccines in the pipeline at GSK Biologicals.
RSV = respiratory syncytial virus; EBV = Epstein–Barr virus; VZV = varicella zoster virus.

sufficient R&D in the vaccine field. It can certainly be said that for the GSK pipeline – and this applies equally to the pipeline of any vaccine-producing company – there is today a vaccine under development for all diseases of the developed and developing worlds, and that complaints of inadequate R&D in the industry are no longer justified and should be reassessed (Figure 6.3). It is now the responsibility of the governments of the developing countries and the G-8 states to provide support, since if vaccine development is to be pursued then there must ultimately be a market for that development to pursue. If this is not the case, then the industry will certainly starve. Of course, the major question today is how will society pay for the 20 new vaccines that will be launched during the next 10 years? This is true not only in developing countries, but also in Europe and the USA. If the value of vaccines is not recognized, then the objective will not be achieved, and the tenet that the control and prevention of infectious disease is better than cure will not be upheld.

6.3
Vaccine Manufacture

Vaccine manufacture is a highly complex process, and it is important that this point is recognized. Producing vaccines was my first task when I joined the vaccine industry, and then as now it remains the most complex task the industry has to offer. In order to produce a modern vaccine, and then to license it, the decision to invest in the manufacturing process must be taken some five years before the launch. On occasion, it may take over a year to produce a vaccine and release it. Typically, during production there will be batch-to-batch deviations, mainly because when working with live microorganisms, deviation is "normal". It must then be determined whether this deviation affects the quality of the vaccine, or not. This problem in its entirety is almost impossible to solve, not only because the system is highly complex but also because the health authorities do not understand such complexity. The authorities will produce a rule-book, but they do not understand the biology behind vaccine manufacture. Twenty years ago, inspections were carried out by biologists, but today's inspections are carried out by people who know the rules but do not understand what they are inspecting. This presents a major problem that must be solved. On the other hand, the industry must bear the responsibility of maintaining up-to-date systems, so that they are not producing vaccines with a plant that was built 20 or 50 years ago. The technology involved is rapidly changing, and the role of the regulators (including WHO) is to indicate when modern technology should be introduced into the process. The role of the regulator is also to understand and finalize a process, and in this respect they must organize and schedule their inspections so that production-line accidents, as have happened during recent years, may be eliminated.

So, the vaccine-producers as a group are requesting an open and constructive relationship with authorities, who in turn should not be afraid of discussing and proposing change, rather than threatening the closure of a facility, as is often the case. For this system to succeed, the main requirement is long-term planning, and recent developments with UNICEF have led to major improvements in this area. As vaccine manufacture is highly complex, it is reasonable to seek three- and five-year forecasts, rather than to operate on a day-to-day basis, and in this way it will be possible to build vaccine stockpiles in case of emergency. Finally, as is the case of the automobile industry where Renault and Volkswagen, for example, have plants in China and India, the vaccine industry will ultimately have plants situated around the world. In this way, it will be feasible for GSK and other vaccine producers to impose the same standards in the developing world as they do in Europe or in the United States. Given the opportunity, such globalization of manufacture could be undertaken very rapidly.

6.4
The Vaccine Marketplace

In the past, the marketing of vaccines often included a launch in Europe and the US, followed some 20 years later by extension of the market into developing countries. This concept must change. A new system is needed in which the vaccine is launched worldwide, in both the private and the public markets. For developing countries, a partnership is needed in order for the industry to continue to invest in vaccine production for the developing world, and support is also needed in areas of R&D. In addition, there is also a need to establish credible markets in these developing countries. As Dr. Martin suggested in Chapter 5, if there is no sustainability, it is likely that the process will not work. It is at this point that the initiative of the British Chancellor, Gordon Brown, should be applauded, in aiming towards a pre-purchase agreement with the vaccine industry.

6.5
Vaccine Availability

I believe that the provision of vaccines free of charge to the entire population is not sustainable for society in general. The main crisis here would be the cost to social security services, as governments are unlikely to continue offering vaccines free of charge to developed countries. The problem is that such a system would have no accountability, and in the public opinion anything that is provided free of charge has no value, and so the system will need to be changed. It is likely that, before 2010, there will be six or seven new vaccines available which Europe is not ready to fund, the United States will probably not fund, and developing countries are simply unable to fund. This will also mean that new vaccines are available for segments of the population other than infants – there will be vaccines for the elderly, for adults, and also for adolescents. Thus, the financial pressure on the community to spend more on disease prevention will increase. Today, only 1.2–1.5% of the global expenditure for pharmaceuticals is spent on vaccination. This percentage must be increased dramatically over the years to come. In terms of vaccine availability, the aim is to provide vaccines to the developed and developing countries, to all segments of the population, and in turn to create market segmentation (Figure 6.4). In India, for example, some people can afford vaccines but do not pay for them, because these are provided for free by UNICEF or WHO. The governments must also take responsibility, and in truth the private and public systems must coexist. In future, there will be an increase in manufacturing capacity, and GSK will clearly invest in all major countries. With plants currently operational in China and India, vaccine manufacture will be continued. And, as with the automobile industry, where needs exist there will be differential pricing. In this way, GSK is looking to the future market and market segmentation, and this must be carried out in cooperation with governments as our company cannot do this alone. The segment at the bottom of Figure 6.4 is termed 'funded'; in other words, a mechanism must be found whereby a wealthy country will help to provide the vaccine for a poorer country (a 'poor' country would be defined as having an annual GDP below 600 dollars per inhabitant), and this is the goal of the

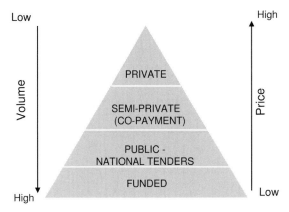

Figure 6.4 Vaccine market segmentation.

Gates Foundation. I applaud what Bill Gates is achieving through his foundation, but as has been already pointed out in Chapter 5, it should not be the case where an individual person provides for half of the funds needed; rather it is the responsibility of the governments of the wealthy countries. To demonstrate what the reality is today, three examples – rotavirus, human papilloma virus, and malaria – will subsequently be examined.

6.5.1
Rotavirus

Rotavirus (Figure 6.5) kills 500 000 children each year, but effective vaccines are becoming available, and at GSK two standards are currently under development. Our first vaccine had been developed in Mexico, and has been launched there at the standard of the U.S. and Europe. Studies were conducted

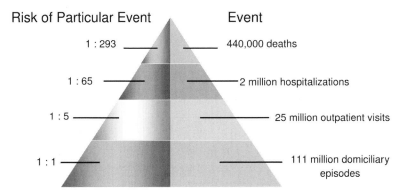

Figure 6.5 Estimated global prevalence of rotavirus disease.
(Source: Parashar et al., Emerg. Infect. Dis. 9, 565–572, 2003).

in 100 000 children – perhaps the first time that a vaccine has been studied in so many children – and the vaccine is now available, though whether it will be used is unclear. To illustrate this, the parameters of the disease must be considered. Rotavirus is a democratic disease, and whether people live in developed or developing countries, they will contract it. However, whereas 1 in 300 children contracting rotavirus will die from it in developing countries, this will not be the case in Europe and the US. Nonetheless, since in Europe and the US 50% of hospitalizations during winter are due to rotavirus, there is a clear medical need, and it will be interesting to see how society implements this vaccine.

6.5.2
Human Papilloma Virus (HPV)

In the treatment of cervical cancer, a vaccine will be launched worldwide in 2007 and, in theory, should be administered to immunize 1.6 billion women. This is the target, and it is the responsibility of government to ensure vaccination. Clearly, a mixture of private and public financing will be required, but it must be remembered that cervical cancer is the second most prevalent cancer in women worldwide. During the past 30 years, no single vaccine has ever been 100% effective, but the HPV vaccine retains 100% efficacy at three, and even perhaps four, years after immunization (Figure 6.6). The medical costs of the disease are huge, and they could be saved by using the vaccine (Figure 6.7). However, it must then be ensured that the savings are reallocated to vaccine purchase, and not used elsewhere. Indeed, this will be the main challenge when the vaccine is launched.

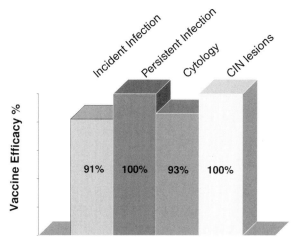

Figure 6.6 Efficacy of the HPV-001 human papilloma virus vaccine. (Source: Harper et al., Lancet 364, 1757–1765, 2004).

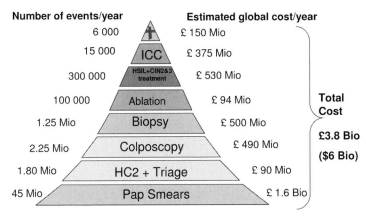

Number of events/year

6 000

15 000

300 000

100 000

1.25 Mio

2.25 Mio

1.80 Mio

45 Mio

ICC

HSIL+CIN2&3
treatment

Ablation

Biopsy

Colposcopy

HC2 + Triage

Pap Smears

Estimated global cost/year

£ 150 Mio

£ 375 Mio

£ 530 Mio

£ 94 Mio

£ 500 Mio

£ 490 Mio

£ 90 Mio

£ 1.6 Bio

**Total
Cost**

£3.8 Bio

($6 Bio)

Figure 6.7 Costs associated with HPV in the United States.

6.5.3
Malaria

The malaria vaccine represents an excellent example of public-private partnership and indeed, without collaboration with Walter Reed Hospital (US army) and the Gates Foundation, its development would not have been possible. The disease is limited to the poorest countries – there is no other market (Figure 6.8). Historically, the development of malarial vaccine extends back 84 years, while the GSK program was started more than 20 years ago, in 1982 (Figure 6.9). The reasons for our sustained effort to develop this vaccine are two-fold. The first reason was to develop a malaria vaccine; the second reason was that malaria has provided various means of improving and increasing scientific knowledge in immunology and infectious disease for the future. Hopefully, the vaccine will be launched in 2009, and will provide about 58% protection against severe disease in the cohort of one to four years. However, when examining the outcome under two years, the protection against severe disease is already 77% (Table 6.1). Now, the focus will be to conduct a major clinical study in children aged between 2 and 18 months, where efficacy of the vaccine would surely be higher and this will have a major impact on public health.

Table 6.1 Results of the RTS,S vaccine trial in Mozambique.
(Source: Alonso et al., Lancet 364, 1411–1420, 2004).

Condition	Mean vaccine efficacy [%]	95% confidence interval
Infection	45	31–56%, $p < 0.001$
Clinical disease	30	11–45%, $p = 0.004$
Severe disease	58	15–79%, $p = 0.019$
Severe disease (first 24 months)	77	27–97%, $p = 0.018$

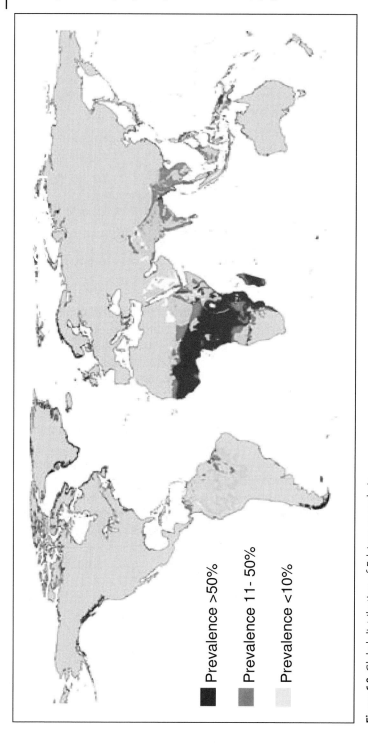

Figure 6.8 Global distribution of Falciparum malaria.
(Source: Snow et al., Nature 434, 214, 2005; Sachs et al., Nature 415, 680, 2002).

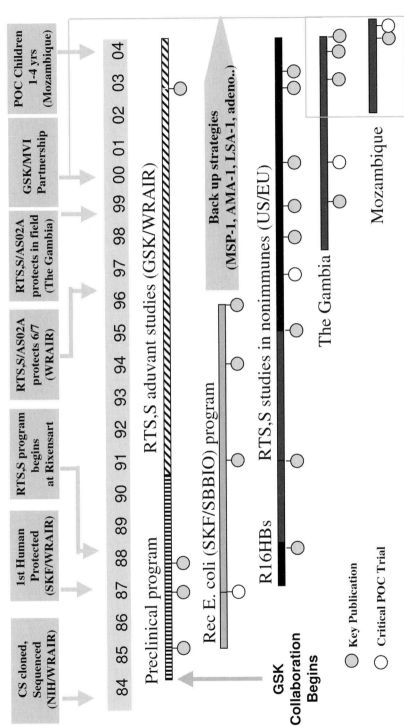

Figure 6.9 History of the RTS,S malaria vaccine program.

6.6
Conclusions

On 28th February 2005, in his budget speech, the Minister of Finance of India stated that, "India is not a poor country. Yet a significant proportion of our people are poor". Whether India is poor is open to discussion, because there are people in India who have money, who can purchase modern vaccines, and who must invest in purchasing vaccines. But there must also be in India a responsibility of the government to purchase certain vaccines for the poorest. If the government accepts this responsibility, then the G-8 group can help them to achieve this. In India, there are three main groups responsible for payment: the wealthy among the Indian population; the federal government; or the local government, who must also pay for the poorest, and this is where the G-8 group can help. This is the preferred type of partnership for vaccination, and is what must be sought. However, if there is no accountability, this approach will fail. The advantage of the Gates Foundation and the Vaccine Fund is that they have created accountability, and it is important that, in India, both the Minister of Finance and the Minister of Health recognize in their budget process that vaccination is important.

In conclusion, it appears the R&D community and the industry have done their job, and now the baton is being passed to others to continue the fight. It must also be realized that vaccine manufacture is a complex process and that industry and the authorities must generate a positive dialogue. If this cannot be achieved, then major crises, as occurred in 2004 for the influenza epidemic, will undoubtedly recur. This is a critical point, because it is necessary not only to change the way that the biomedical industry functions but also to ensure an ongoing dialogue to ensure increased output and quality control on a worldwide basis. Also under debate is whether the value of vaccines will be recognized in each country. This relates to both developing and developed countries alike, and in future vaccines should be regarded as an investment, similar to that of education, in order for society to progress. The final question is whether the G-8 powers and other governments will apply the new model to the cases for rotavirus, HPV, and malaria. This is the next major challenge, and it seems certain that the model will be applied also to streptococcus pneumoniae, and perhaps later for dengue fever. This is the challenge, and it is here now.

General Bibliography and Suggested Reading

1 P. A. MUSGROVE (Ed.), *Health Economics in Development* (**2003**), World Bank publication.

2 T. ACHARYA, R. KENNEDY, A. S. DAAR, Biotechnology to improve health in developing countries: a review (**2004**), *Mem. Inst. Oswaldo Cruz* 99, 341–350.

3 Betterhumans website, Betterhumans Staff, *Diseases of Developing World the Target of Biotech Initiatives.*
http://www.betterhumans.com/News/news.aspx?articleID=2003-01-29-2.

4 Genome Canada website, *Experts rank top 10 biotechnologies for improving global health within 5–10 years.* http://www.genomecanada.ca/GCmedia/communiquesPresse/indexDetails.asp?id=133&l=e.

5 Global exchange website, S. HURLICH, *The World's First Synthetic Vaccine for Children: The Cuban Face of Biotechnology.*
http://www.globalexchange.org/countries/cuba/foodAndMeds/1510.html.

6 M. MOTARI, U. QUACH, H. THORSTEINSDOTTIR, D. K. MARTIN, A. D. DAAR, P. A. SINGER, South Africa – blazing a trail for African biotechnology (**2004**), *Nat. Biotechnol.* 22, Suppl. DC37–DC41.

7 P. A. SINGER, A. S. DAAR, Harnessing Genomics and Biotechnology to Improve Global Health Equity (**2001**), *Science* 294, 87–89.

8 A. P. WATERS, M. M. MOTA, M. R. VAN DIJK, C. J. JANSE, Malaria Vaccines: Back to the Future? (**2005**), *Science* 307, 528–530.

9 M. ADEL, The Global Vaccination Gap (**2004**), *Science* 305, 147.

10 R. S. DESOWITZ, *The malaria caspers: mores tales of parasites and people, research and reality* (**1993**), W. W. Norton Company (reprint).

11 J. L. DI FABIO, C. DE QUADROS, Considerations for combination vaccine development and use in the developing world (**2001**), *Clin. Infect. Dis.* 33, Suppl. 4, S340–S345.

12 R. T. MAHONEY, J. MAYNARD, Introduction of new vaccines into developing countries (**1999**), *Vaccine* 17, 646–652.

13 R. T. MAHONEY, S. RAMACHANDRAN, Z. XU, The introduction of new vaccines into developing countries II. Vaccine financing (**2000**), *Vaccine* 18, 2625–2635.

Health for All?: Analyses and Recommendations
Edited by The World Life Sciences Forum – BioVision
Copyright © 2005 Wiley-VCH Verlag GmbH & Co. KGaA, Weinheim
ISBN: 3-527-31489-X

14 R. T. MAHONEY, A. PABLOS-MENDEZ, S. RAMACHANDRAN,
The introduction of new vaccines into developing countries.
III. The role of intellectual property (**2004**), *Vaccine* 22, 786–792.

15 W. A. MURASKIN, *The politics of international health: the children's vaccine
initiative and the struggle to develop vaccines for the Third World* (**1998**),
State University of New York Press.

16 B. A. WEISBROD, J. H. HUSTON, *Benefits and costs of human vaccines
in developed countries: an evaluative survey*, 2nd edition (**1983**),
Pharmaceutical Manufacturers Association, Washington, D.C.

17 P. F. BASH, *Vaccines and world health: science, policy, and practice* (**1994**),
Oxford University Press, New York.

18 B. BLOOM, P. H. LAMBERT, *The vaccine book* (**2002**), Academic Press.

19 L. GALAMBOS, J. ELIOT SEWELL, *Network of innovation: vaccine development
at Merck, Sharp and Dohme, and Mulfords, 1895–1995* (**1996**), Cambridge
University Press.

20 R. RODRIGUEZ-MONGUIO, J. ROVIRA, *An analysis of pharmaceutical
lending by the World Bank.*
http://www-wds.worldbank.org/servlet/WDSContentServer/WDSP/
IB/2005/02/23/000090341_20050223150015/Rendered/PDF/
315100HNP0Phar1ng0Analysis01public1.pdf.

21 PhRMA (Pharmaceutical Research and Manufacturers of America)
website, *Why do medicines cost so much,* 1. http://www.phrma.org/
publications/publications/brochure/questions/whycostmuch.cfm.

22 S. A. PLOTKIN, W. A. ORENSTEIN, P. A. OFFIT, *Vaccines*, 4th edition
(**2003**), W. B. Saunders Company.

23 United States National Library of Medicine website, "Multilateral
Initiative on Malaria" Builds Unique Coalition to Aid African Scientists
(**2000**), *NLM Newsline* 55 (4). http://www.nlm.nih.gov/archive/
20040423/pubs/nlmnews/octdec00/Malaria.html.

24 L. WROUGHTON, World Bank to Expand Fight Against Malaria,
Reuters Health, April 25, **2005**.
http://www.nlm.nih.gov/medlineplus/news/fullstory_24280.html.

25 Voice of America website, D. McALARY, Group Urges Drug Companies
to Make Vaccines for Poor Countries, *Voice of America News*, April 8,
2005. http://www.voanews.com/english/2005-04-08-voa69.cf.

26 SOHU website, Channelling more funds to the poor, *China Daily*,
February 17, **2005**.
http://english.sohu.com/20050217/n224318018.shtml.

27 Aids Education Global Information System website, Fact Sheet:
New Partnerships to Develop and Deliver Vaccines for Diseases
(HIV/AIDS, Malaria, and Tuberculosis vaccines for developing
countries), *Washington File*, March 2, **2000**.
http://www.aegis.com/news/usis/2000/US000302.html.

28 UNICEF website, UNICEF Press release, Business-like approach to funding health programs in poor countries may save more than two million lives in 5 years.
http://www.unicef.org/media/media_18857.html.

29 A. ROBBINS, I. ARITA, The global capacity for manufacturing vaccines. Prospects for competition and collaboration among producers in the next decade (**1994**), *Int J. Technol Assess Health Care* 10 (1), 39–46.

30 U.S. Food and Drug Administration Website, Facilitating Efficiency in Vaccine Manufacture: *Helping Defeat the Scourge of Meningococcal Disease in Africa.* http://www.fda.gov/oc/initiatives/criticalpath/vaccine.html.

31 Europaworld website, Affordable Global Vaccine Supply At Risk, October 25, **2002**.
http://www.europaworld.org/week102/affordableglobalvaccine51002.htm.

32 Crucell Press Release, Crucell and NIH Sign 21.4 Million Ebola Vaccine Manufacturing Contract, April 14, **2005**.
http://hugin.info/132631/R/989152/148350.pdf.

33 Biospectrum website, Affordable Immunity With Superior Technology, Indian Immunologicals, October 05, **2004**.
http://www.biospectrumindia.com/content/movers/10410051.asp.

34 News-Medical.Net website, Flu vaccine shortage highlights need to accelerate the approval of affordable generic biopharmaceuticals and vaccines, *News-Medical.Net*, October 8, **2004**.
http://www.news-medical.net/?id=5431.

Module III
Bioethics: What is the Tradeoff Between Principles and People?

Health for All?: Analyses and Recommendations
Edited by The World Life Sciences Forum – BioVision
Copyright © 2005 Wiley-VCH Verlag GmbH & Co. KGaA, Weinheim
ISBN: 3-527-31489-X

Introduction

Dominique Lecourt

The success of innovations in vaccines rests on the technological capacity of innovation, the productivity of the vaccine industry, and the conquest of new markets. However, medical treatment of the human body set in highly diverse cultural contexts necessitates careful attention to ethics. In the case of sub-Saharan Africa this point of view merits even closer scrutiny. The progress of clinical trials which proceed there must be followed rigorously to ensure the enlightened assent of the subjects which is freely obtained, and that the populations concerned will be able to profit from any favourable result. But how far should we go in promoting this regulation? Companies are increasingly risk-averse, and are likely to slow the development of vaccines in keeping with increased regulatory oversight. Once again the commitment of politicians is as important as that of economic decision makers. They must ensure that the general public is able to advocate for their own health by including, understanding and supporting the efforts of scientists and the industrialists.

Health for All?: Analyses and Recommendations
Edited by The World Life Sciences Forum – BioVision
Copyright © 2005 Wiley-VCH Verlag GmbH & Co. KGaA, Weinheim
ISBN: 3-527-31489-X

Author Biography

Philippe Kourilsky

Director, Pasteur Institute (France)

Philippe Kourilsky received a doctorate from l'Ecole Polytechnique. In 1972, he joined the Institut Pasteur serving as the director of the Molecular Biology Unit in 1979 and as the director of research from 1992 to 1995. He was named a professor at the Institute in 1993, and in 1998 he was also named a professor at the College de France where he holds the chair of Molecular Immunology. In 2000, he became the general director of the Institut Pasteur. He is a member of the French Academy of Sciences.

7
Are First-World Ethics Applicable to the Third World?

Philippe Kourilsky

7.1
Introduction

In human health, less than 10% of R&D resources are devoted to diseases that involve 90% of the world's population. More than two billion people suffer from infectious diseases for which vaccines and drugs are not available, because they are too expensive, because they cannot be adequately distributed, or because they simply do not exist. The R&D pipelines which could – and should – provide adequate vaccines and drugs for such neglected diseases, face major difficulties. Contrary to common thinking, funding is not the only issue, as research, regulations and ethics each share in the unacceptable situation endured by almost half of the world's population. In fact, the numbers of disadvantaged are absolutely huge, numbering billions among the human population.

7.2
The "90–10 Gap"

The problems faced today are double-faced. First, there is the problem of distributing existing vaccines and drugs, and it is quite well-known – and incredibly scandalous – that the measles vaccine which costs only a few cents and has used for decades is still not distributed to the extent that several hundred thousand children die every year. The second issue is that of R&D. This problem is referred to as the "90–10 gap", which means that 90% of disadvantaged people benefit from less than 10% of the worldwide biochemical and biomedical research (Figure 7.1).

Effort in terms of R&D is very much focused on the diseases that affect the developed world; for example, much more money is spent on prostate cancer

Health for All?: Analyses and Recommendations
Edited by The World Life Sciences Forum – BioVision
Copyright © 2005 Wiley-VCH Verlag GmbH & Co. KGaA, Weinheim
ISBN: 3-527-31489-X

> • **90 % of the diseased people benefit from less than 10 % of the worldwide biomedical research**
>
> • **1 % of the 1400 new drugs which have reached the market in the last 25 years are devoted to the socalled "neglected diseases"**

Figure 7.1 The "90–10 gap".

than on tuberculosis. As a result, only 1% of the 1400 new drugs that have reached the market during the past 25 years are devoted to the so-called neglected diseases – that is, diseases of the poor. Consequently, R&D is not providing the medical solutions for very large numbers of people. Thus, it is not only the correct distribution of existing products in the developing world that is lacking, but also several products needed to cure diseases in the developing world. It must be emphasized that, during the past 25 years, worldwide funding in R&D has increased at least three- to five-fold, and on that basis it cannot be argued that there is insufficient money available. Rather, it is the way in which that money has been distributed which is in question. In other words, there is a structural problem that needs to be understood.

Research is to a significant extent (but not entirely so) market-independent, whereas development is to a large extent market-dependent. In other words, academia is virtually free to do what it wants in research, and in areas that it wants to explore. Governments provide a very strong stimulus when allocating resources within academia for a certain type of investigation. The question should also be raised with regard to the role of charitable money in the promotion of research; for example, much more charity money is donated for research into cancer than into infectious disease. However, research is basically or largely market-independent and in academia – for example, in the Pasteur Institute and elsewhere – there is freedom to work on diseases for which there is no drug market. This is not the case for development which is, to a very large extent, market-dependent. Consequently, the problem is one of how to finance studies, with development costs being at least ten-fold those of the research. In today's world, the development costs of a new drug are estimated to be approximately 500 million Euro. Likewise, the costs to develop a vaccine have increased significantly, and are now similar to those for a drug, with huge amounts of money being spent in order to reach the marketplace. This is especially problematic for the neglected diseases, because if there is no marketplace then there is no new medicine. The research may be conducted, but there are no drugs or vaccines to follow. At the Pasteur Institute, a large proportion (up to half) of the research effort – even if successful – will never be developed into real products usable in the developing world, because development funds are lacking. The basic problem is that, in an ongoing research program, development will not fall within the current funding system and products will not be generated. For the neglected diseases this represents a major disaster, the conclusion being that research is neglected to a certain extent, notably in areas of exotic parasitism, such as filariosis.

7.3
Vaccine Production

Vaccines are a neglected part of the pharmaceutical industry in the sense that they represent less than 3% of the total drug market. If a projection is made based on the fact that vaccine-producing companies spend exactly the same fraction of their money on R&D, then the industrial money available for vaccine development should be about 3% of what is available to develop drugs. Put this way, it may not be surprising that, as yet, there are no vaccines available to combat HIV while antiretroviral drugs have, fortunately, been made available almost ten years ago. Although it would appear that the funds made available for HIV vaccine development are insufficient, this is not accepted by all, and it is also true that an HIV vaccine involves specific and difficult research problems that must be addressed.

Until recently, the manufacture of products resulting from R&D projects was mainly carried out in the developed countries. The cost of the product incorporates the cost of R&D, which means that the cost structure is modeled by the developed countries and not by the developing countries. This cost structure is clearly inadequate for developing countries, however. A major point here is the impact of regulatory standards. It is acknowledged by the major pharmaceutical companies that development costs have increased three-fold during the past 15 years, and this is especially the case for vaccines. The question to be asked, then, is whether this increase is justified, and why. In fact, the rise in costs is mainly linked to a demand for increased safety among the general public, and is reflected in many processes. Today, for example, liability problems in developed countries have become very significant, with companies withdrawing drugs as soon as any safety problem appears. Often, the withdrawal is upheld even when the situation is manageable, for fear of the company being sued and losing very large amounts of money.

The consequences of this for the developing countries are numerous. For example, as standards change almost daily a "race" is developing that they cannot keep up with. The regulatory agencies seem to compete with each other, with the European agency, the EMEA, wanting to perform as many good deeds as the FDA. As new technologies are developed, it is temptingly normal to introduce new analytical methods in an attempt to achieve improved purity, characterization, etc. The problem is that the developing countries cannot maintain the pace. Accordingly, they face a highly efficient protectionist barrier against the export of health products to the industrialized countries of the North. This is very different from buying a shirt or a suit, many of which are now produced in China, the point being that the drug industries of the western world are very well protected by their regulatory standards.

7.4
The Ethics of Regulatory Standards

One particular issue is that when paying for the drugs and vaccines manufactured in the developed world, the developing countries suffer from inadequate cost structure. In other words, they pay for the R&D and for manufacturing performed along the standards of the rich countries. This can be partly corrected by the differential pricing policies being set forth, but this is a relatively recent proposal, except for vaccines. The developing countries often apply the same standards to themselves because they do not want to appear inferior to the rest of the world. As a result, they increase their own problem-solving approaches, and this may have dire final consequences. The R&D costs for vaccines and drugs that are specific to neglected diseases are beyond the available means of all. In fact, they are beyond the available means of the countries in which these products are needed, as well those of most nongovernmental organizations (NGOs) and of the international organizations themselves. Put another way, if 50 products are needed, and each requires about US$ 1 billion for development, the money is simply not available worldwide. One point that should be emphasized here is the belief that many people have, that regulatory standards fit the ethic. In other words, there is an implicit ethical dimension in regulatory standards, which can be stated in very simple terms that "safer is better". In general, people do believe that safer is better, and that this is the ethical statement. Today, however, it is unclear to what extent safer is obviously better, and how such a statement can be introduced into a cost benefit analysis which has been adapted to the real world.

In the modern, global world there is a clear trend towards the globalization of standards, with pressure being applied constantly to achieve this goal. There are very many reasons for this proposal, some are technical, and others political. The ethical movements, or at least part of the ethics thinking in the world, is termed "universalist", and this now claims that the ethical principles should be worldwide. Although it is unquestionable that health standards should be the same in all parts of the world – and the achievement of this goal would be excellent – it is so far away as to be totally unrealistic. Indeed, it could be considered unethical to imagine that this might be the case in the short term.

7.5
Challenging Regulatory Standards

One final question relates to whether regulatory standards should be challenged. The most likely answer to this is "yes", because the benefits and costs incurred are poorly evaluated (if at all), and this is especially shocking for scientists. It is amazing to see that these regulations have varied over time, and that the costs have multiplied by a factor of two to three, and yet there is no evaluation. In the case of vaccines, most people living today have been treated with vaccines that today would most likely be claimed as inadequate, because they would not match the standards. For example, BCG is a very old vaccine that is known not to be perfect, and it is accepted that a better tuberculosis vaccine is needed. The BCG vaccine provokes a very small number of accidents termed BCG-itis in children, and consequently the claim has been made that the vaccine should be improved, or better purified. In fact, the cohort of children who contract BCG-itis has been studied, with remarkable results. All children who developed the condition were found to have mutations in the immune system which were revealed by vaccination. In a way, they were immunodeficient, but this had not been noticed because it was only a slight immunodeficiency that was revealed by the vaccination. To put this another way, the safety of the vaccine cannot be improved; the problem is not due to the vaccine's characteristics, but simply to genetic variation among the human population.

It is thus clear that the regulatory standards must be challenged in order to determine why the costs are so high, and what benefits are obtained. This is especially important for the developing countries, as elsewhere in the world the cost of health is growing, and all governments will attempt to limit the public costs. Even in the developed countries it is important to determine how and when the money is spent, since the social security systems are limited in the amounts they can pay. Challenging the regulatory standards does not imply any relaxation in scientific rigor; on the contrary, it provides a better indication of what is being done and whether it is being done properly.

Surprisingly enough, regulatory standards have their share of responsibility in unacceptable health situations in the developing world. R&D costs cannot be borne by poor countries, academic institutions, NGOs, or even international organizations. The budget of the Global Alliance for Vaccines and Immunization (GAVI), which is 1.5 billion US$ over a five-year period, would allow for the development of two molecules or two vaccines. The issue is whether such funding is sufficient. A recent directive involved the creation of a new NGO termed the Drugs for Neglected Disease Initiative (DNDi), the business plan of which was created by Doctors Without Borders, together with the Pasteur Institute and various other organizations. The business plan is to produce eight new molecules for infectious diseases which are completely neglected, an example being Chagas' disease. The initial budget for this is 50 million US$, received mainly from donations. Of course, the DNDi is seeking more money, but if eight molecules each costing 500 million US$ are required, this clearly is an impossible task. The main objective, therefore, is to determine how to develop drugs at lower cost, whilst maintaining the same level of safety. It seems strange that the public's generosity should be misspent on processes of unproven usefulness.

7.6
Future Strategies

What, then, might be the answer to this problem? What strategies should be followed? The first stage would be a scientific cost-benefit analysis of regulatory standards, and this should be strongly promoted. In addition, the local structures should be strengthened and bottom-up approaches built (Figure 7.2), as is the case for the international network of the Pasteur Institute. The Institute is an historical legacy of Louis Pasteur himself, with several institutes having been introduced at the end of the nineteenth century and located worldwide. Strikingly, whilst only 21 institutes were in existence four years ago, the current number is 29 (Figure 7.3). Requests are made throughout the world to share problems of public health, and this is carried out using a bottom-line approach in which it is important to perceive problems and integrate them from the ground up, rather than the other way round. An example taken from the network illustrates that the issues under discussion are not theoretical. Traditionally, the Pasteur Institute in Cambodia vaccinates freely against rabies, as the condition there is widespread among infected dogs. Each year, the Institute receives some 12 000 people who have been bitten by dogs. The vaccine used was originally produced by the Pasteur Institute in Vietnam, using a very old method that employed mouse brain.

➢ Promote a scientific cost-benefit analysis of regulatory standards

➢ Strengthen local structures and build bottom-up approaches

➢ Support innovative partnerships

➢ Raise awareness and generosity

Figure 7.2 Strategies that should be followed.

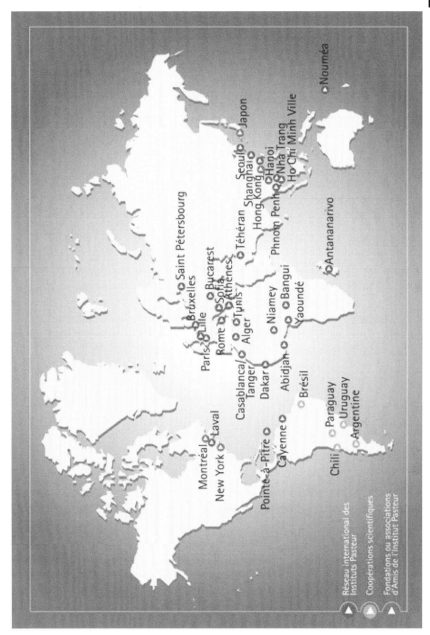

Figure 7.3 Worldwide locations of the Institut Pasteur in 2005.

The WHO has, perfectly correctly, recommended that this procedure be stopped, but the new vaccine, which is made on Vero cells, costs five- to ten-fold more than the original. The Pasteur Institute in Cambodia could not afford to buy the vaccine, and so the vaccination rate has plunged. There had been no accidents with the previous vaccine, but the WHO recommendation was clearly correct. Fortunately, friendship and good relations with the industry led to vaccine donations and a transient solution of the problem. Nonetheless, this is typical of the difficulties faced in practice.

Among health initiatives, the GAVI performs an excellent role, while other partnerships such as DNDi, the Medicine For Malaria Venture, and the Global Alliance for Tuberculosis each also deserve support. It is important also to raise awareness and generosity. Awareness means returning to the facts and numbers (as unpleasant as they may be), and accepting the questions. For example, in the United States, a major effort has been made to develop a new vaccine against bioterrorist-transmitted smallpox. In this case, emergency procedures for regulatory processes and development were used to accelerate vaccine production. Why this has not been considered for AIDS, which is an enormous problem? In France, on the other hand, live vaccines are considered as genetically modified organisms (GMOs), with special regulations for their handling having been introduced, for environmental reasons. Sadly, one vaccine candidate was kept in cold storage for a year because the committee involved with the vaccine's environmental aspects took that time to realize that it did not pose any such problem.

Perhaps the question should be asked again – what is the trade-off between principles and people? These are practical issues which must be considered in order to develop the theoretical thinking that supports the action taken.

Author Biography

Hoosen Mohamed Coovadia

Victor Daitz Chair in HIV/AIDS Research, Nelson R. Mandela School of Medicine,
University of KwaZulu-Natal (South Africa)

Hoosen Mohamed (Jerry) Coovadia has distinguished himself over many years as a leading paediatric immunologist, a leader in the struggle for a democratic South Africa, a national and international figure in the paediatric world and, more recently, a world authority in the field of paediatric HIV/AIDS, both as a researcher and as a powerful force in shaping policy with respect to the disease.

He specialised in paediatrics at the University of Natal and became a Fellow of the College of Paediatricians of the College of Medicine of South Africa in 1971. In 1974, he obtained his MSc in Immunology from the University of Birmingham. Returning to South Africa after his studies in the UK, he rejoined the Department of Paediatrics at the University of Natal and began to work on the immunology of measles in children. His research in that field led to the award of an MD in 1978, the year in which he was appointed Principal Paediatrician and Senior Lecturer. In 1982 he was appointed Associate Professor and in 1986 Ad Hominem Professor. In 1990, he became Professor and Head of Paediatrics and Child Health at the University of Natal, until the end of 2000. During that time, he created a strong and vibrant department held in high regard for its teaching, clinical excellence and research.

After retiring from this position, H. M. Coovadia was appointed the Victor Daitz Chair in HIV/AIDS Research, and Director of Biomedical Science at the Centre for HIV/AIDS Networking at the Nelson R. Mandela School of Medicine, University of Natal.

His interest in paediatric HIV/AIDS developed in the early 1990s as the extent of the tragedy in South Africa began to be recognised. His particular interest has the transmission of the virus from mother to child. Over the years he has attracted numerous large research grants from both local and overseas donors and has built up a powerful research team at the Nelson R. Mandela School of Medicine.

H. M. Coovadia, with his research group, has published a number of ground breaking research articles. They were first to suggest that, contrary to received opinion, transmission of HIV from mother-to-child via breast feeding might be substantially reduced if the mother exclusively breastfeeds. While research in this area is not conclusive, if the theory is confirmed it will have significant impact on the health of mothers and children, particularly in developing countries, where formula feeding carries great dangers.

He was appointed by the National Department of Health as Chairperson of the National Advisory Group on the HIV/AIDS and STD Programme from 1995 to 1997, while his international stature in the area of HIV/AIDS led to his election as Chairperson of the XIIIth International Conference on AIDS, held in Durban in July 2000.

Among the other national leadership positions held by H. M. Coovadia over the years are those of Deputy Chairperson of the Transitional National Development Trust of South Africa, Trustee of Independent Development Trust, Chairperson of the Commission on Maternal and Child Health Policy set up by the government in 1994. He is also a founder member of the South African Academy of Sciences.

His research output is prodigious – he has authored or co-authored more than 200 articles in peer reviewed journals, many of them leading international journals. He is co-editor of the textbook Paediatrics and Child Health, which is widely used by medical students and junior doctors throughout South Africa.

H. M. Coovadia has received numerous accolades and awards. He was elected a Fellow of the University of Natal in 1995 and was awarded an honorary DSc by the University of Durban Westville in 1996. In 1999 President Nelson R. Mandela honoured him with the Star of South Africa for his contribution to democracy and health and he received a silver medal from the Medical Research Council for excellence in research. In 2000 he received the International Association of Physicians in AIDS and Care Award, the Heroes in Medicine Award in Toronto, Canada, the Nelson Mandela Award for Health and Human Rights and he was elected a Foreign Member of the Institute of Medicine of the National Academy of Sciences.

8
Does Biotechnology Serve Africa's Needs?

Hoosen Mohamed Coovadia

8.1
Introduction

AIDS lends itself to a sense of melodrama, and endless statistics of doom. South Africa has the largest number of HIV/AIDS cases worldwide, and its society – which ironically, when the condition first arose, was just beginning to enjoy freedom – has been devastated by this epidemic. The condition reflects, in its most acute and stark forms, some of the social, economic and political disasters that face the African nations, and in particular those which are currently dealing with this HIV epidemic. The lessons learned are not only biological and medical – many are also political, economical, or social.

HIV/AIDS is a disease like no other, and technology is absolutely critical in the fight against this condition. In the past, many technological disasters have befallen Africa, perhaps most notably the disastrous dumping of formula milk. Africa has a huge backlog to catch up, and there are serious measures to try to overcome some of the problems of what appears to be a benighted continent – the problems of governance, of corruption, and of endless wars. Although not finalized, moves are afoot to persuade black governments to become accountable to some authorities. Today, in Johannesburg, there is an institution called the "African Union", and there are a number of institutional maneuvers and efforts at achieving peace. To this end, South Africa is playing a leading role in providing a coherence to Africa, to assist in its development and advancement, that has never previously been attempted. It is fair to suggest that it may well be time for the African continent to begin using biotechnology in a way as never before.

Health for All?: Analyses and Recommendations
Edited by The World Life Sciences Forum – BioVision
Copyright © 2005 Wiley-VCH Verlag GmbH & Co. KGaA, Weinheim
ISBN: 3-527-31489-X

8.2
Biotechnology in Africa

In Africa, available biotechnology is affordable, effective and generally accessible, but is simply not being used. Among the major risk factors for the global burden of disease, malnutrition remains the leading contender (Table 8.1). The need for biotechnology to solve the problem of malnutrition is absolute. It was thought that when freedom and democracy were obtained, then malnutrition would disappear, but this has not been the case. In fact, the situation has worsened, mainly because of the AIDS crisis. It is quite clear that there is a need for both agricultural improvements and biotechnology to provide the economic growth that will solve the problems of malnutrition. Indeed, malnutrition is clearly dependent on the input of biotechnology.

Other risk factors for disease include tobacco smoking, physical inactivity and alcohol intake, and most of these have been well-recognized for decades. In the case of tobacco smoking it has taken about 20 years to re-educate people and to change their smoking habits. However, attempting to alter people's sexual activity as a means of halting the advance of AIDS is much more personal, and any re-education must be carried out with a great deal of sensitivity and persuasion.

The data in Figure 8.1 support the point that has already been made. It is known that there are biotechnologies available which work, and are affordable, and it is also known that the absence of these biotechnologies are a major

Table 8.1 Risk factors and the attributable global burden of disease and injury. (Source: Murray and Lopez, Science 274, 1593–1594, 1996).

Risk factor	*Percentage of total deaths*	*Percentage of total disability adjusted life years (DALY)*
Malnutrition	11.7	15.9
Tobacco	6.0	2.6
Hypertension	5.8	1.4
Poor water supply, sanitation and hygiene	5.3	6.8
Physical inactivity	3.9	1.0
Unsafe sex	2.2	3.5
Occupation	2.2	2.7
Alcohol	1.5	3.5
Air pollution	1.1	0.5
Illicit drugs	0.2	0.6

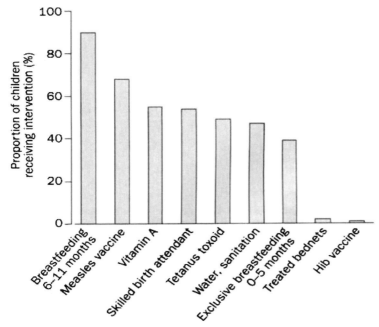

Figure 8.1 Estimated proportion of children aged less than 5 years
who receive survival interventions.
(Data are for 42 countries accounting for 90% of all deaths of children aged
≤ 5 years in the year 2000. Source: Bryce et al., Lancet 362, 159–164, 2003).

cause of mortality in children aged under 5 years. In fact, the under 5-year mortality rate represents one of the best indicators of child health, and of the health services of the country as a whole. Few of the interventions shown in the figure are either unavailable, prohibitively expensive, or unavailable to every child in every developing country, including breastfeeding, measles vaccines and vitamin A intake. For vitamin A, recent reports from the United Kingdom and Nepal have referred to the use of multivitamins to reduce low birth weight. Other reports from Tanzania have shown that simple multivitamin preparations, when given to women during pregnancy and after delivery, can improve reproductive health outcome. Overall, these are simple technologies, and include simple facilities such as skilled birth attendants, tetanus toxoid, water and sanitation, exclusive breastfeeding, treated bed nets and hepatitis B (Hib) vaccine.

Not all of these major interventions require biotechnology, however. Research into any of these issues is very expensive; for example, one study to examine a hypothesis concerning a link between exclusive breastfeeding and HIV has, to date, cost almost 1 million US$. These problems appear simple, but the research and technology involved to solve them is very expensive.

Nonetheless, a recent Institute of Medicine report suggested that one of the simplest, most affordable and most essential technologies (if it can be called that) to reduce neonatal mortality is a skilled birth attendant!

Another recent report has suggested that perinatal or neonatal mortality is the single biggest cause of death among children aged less than 5 years. More importantly, three million of the four million deaths involved could be prevented by the use of simple technologies, such as pre-conceptual folic acid. During pregnancy, these technologies would include syphilis testing, tetanus toxoid, or the treatment of asymptomatic bacteriuria. Ironically, the only expensive component that would increase infant survival to any great degree is that of clinical care of the baby when delivered.

It is impossible to speak for the whole of Africa, because it is such a heterogeneous continent. It has heterogeneous HIV infection rates, it has different behaviors, and it has different rates of sexually transmitted diseases (STDs). However, it is clear that there are very few biotechnology companies in Africa as a whole, and the large majority of these – about 100 in total – are located in South Africa (Table 8.2), though only 47 of these were "core" biotech companies.

One interesting point is that of information technology availability in Africa. With regard to Internet usage, as a continent Africa uses only 1% of worldwide Internet technology, and about 80% of that 1% is in South Africa. There is, therefore, a huge gap to be filled. Clearly, the Internet is another technology which is required, is affordable, and is essential, but is not yet available in Africa.

Table 8.2 Biotechnology resources in the US and in South Africa. (Source: Motari et al., Nature Biotechnol. 22, DC37–41, 2004).

Country	Biotechnology resources
USA	No. of public biotech companies: 300 Market capitalization: US$ 353 billion Annual turnover: US$ 22 billion
South Africa	No. of biotech companies: 106 (of which 47 are "core" biotech)

8.3
The March of AIDS

Many of today's problems in most African countries have arisen as a result of the AIDS epidemic. This is a devastating disease which affects almost every aspect of society. From an economic standpoint, AIDS causes so much sickness among workers that the economy is failing, and this is having a huge impact on the corporate sector. It is vital, therefore, that antiretroviral drugs are provided urgently, and in many respects it seems that South Africa, as a wealthy country, can afford to do this. Working together, the economists and demographers should be able to provide an essential service to help deal with this epidemic.

Despite these suggestions, it is clear that AIDS has proven to be a dreadful epidemic, and not without effect on the government of South Africa. Since the mid-1990s, the scientific community of South Africa has passed through an absolutely excoriating period, there being wide gulfs between what was scientific truth, what was the government's opinion, and what was everyone else's opinion. It has been proposed that, in this world, there is a "democratization of knowledge" – as if every point of view has equal importance. In other words, truth (or something thought to be the truth) can be established scientifically at high financial cost and effort, only to be nullified by the opinion of a politician or the history of tradition.

People entrapped in the HIV epidemic, and without access to antiretroviral therapy, will seek help in whatever way they can. By this stage they are so desperate that they will accept all of these so-called nonscientific parallel medicines that in time will ruin their lives. But this is simply a measure of their desperation, and it is dreadfully wrong.

The point to be made here is that the government of South Africa has played a peculiar role in denying the existence of HIV/AIDS, with the so-called "panel of experts" being drawn mainly from the United States or from European countries where the condition, although problematic, is far less of a national concern (Table 8.3). It appears that the South African government allowed this situation to develop and has continually undermined scientific efforts to deal with the HIV/AIDS epidemic. Since early 1997, the country has been immersed in a series of exaggerated claims for the drug treatment of AIDS. An example of this was virodene; this was developed in South Africa and considered to be effective against AIDS, but the claims made for its success were found to be totally false. Subsequently, nevirapine – which has saved many thousands of children from being infected with AIDS – was later labeled as a poison or toxin. Dangers were broadcast relating to false evidence having been derived from these studies, and claims were made of American imperialism using Africans as "guinea pigs", but this simply was not true.

Table 8.3 Political evolution of AIDS in South Africa.

Year	Event
1996	Sarafina II play performed
1997	Virodene used in AIDS treatment Advisory Committee established
1998	Medical Control Council founded
1999	Zidovudine treatment available Mother-to-child transmission control taken up
2000	"Panel of Experts" established National AIDS Council founded
2001	Mortality data
2002	Global Fund created

The AIDS epidemic and the difficulties associated with it have affected not only the sickness, health and commerce of the people – it has also perverted the democracy of the country and the peoples' freedom because they cannot contradict the authorities. South Africa is a genuine democracy, with multiple foci of power, with scientists who are genuinely fearless to state their cases, and an excellent judiciary system. It would seem that South African society is capable of dealing with such a situation, even if it is unusual. Currently, the money needed to care and treat HIV/AIDS each year in Africa is about 9 billion US$, and this is what the Global Fund aims to achieve (Figure 8.2).

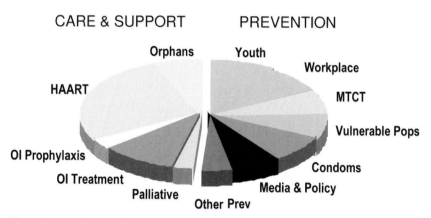

Figure 8.2 Distribution of resources needed for AIDS prevention, care and support. Total funds needed: 9.2 billion US$.

However, these problems simply cannot be dealt with in the absence of an appropriate technology, appropriate drugs, appropriate community mobilizations, and appropriate international collaborations. Indeed, one of the most gratifying issues concerning HIV/AIDS is the large number of organizations that have joined in this task, including the Global Fund, the World Bank fund, and PEPFA funds and the Clinton Foundation's efforts to provide drugs at reduced prices.

In the Kwazulu-Natal province alone, which has a population of about eight million people, the number of international agencies present is almost uncountable. However, nobody seems to know what these people are doing; they may be working at cross-purposes or even repeating each others' investigations, but there appears to be no correspondence between them. But whatever mistakes they are making, the overall cost each year is 9 billion US$. This is a huge amount of money – indeed, it is an amount that most countries in Africa could not afford, except perhaps for South Africa.

8.4
Antiretroviral Drugs

The future approach is much more refined, and antiretroviral treatment will be attempted by using the new "3 by 5" initiative (Figure 8.3). Although this scheme is designed to cover three million people, the need is probably for about 9 to 10 million people. There is, therefore, an enormous cost in terms of human resources, human effort, government commitment and all such other resources required to overcome these gaps. The unfortunate point here is that the percentage of people who need antiretrovirals but who actually receive them is pitifully small (Table 8.4).

The introduction of antiretrovirals into the African nations is critical, and fills the medical community with some trepidation. The situation is less bleak for other conditions; for example, when the Hib vaccine was released, the evidence for its efficacy was so good that it could not be ignored. Likewise, for the pneumococcal conjugate vaccines the evidence was absolutely convincing, with benefits not only of pneumonia reduction but also for child health in general. When recommending antiretrovirals, however, the future is much less secure, and what might happen in another ten years' time is far from clear. The scale of the problem is simply so huge that is extremely difficult to predict what might happen to the country's health services, economy, budgets, and health resources. The reason for such complexity is that AIDS requires the re-training of health workers on a scale never seen before. Likewise, from the patients' angle, it takes only a few hours to treat a child with chronic renal failure, but for a child with AIDS the timescale is simply stupefying. In a province of eight million people, 35% of the women are HIV-infected, which

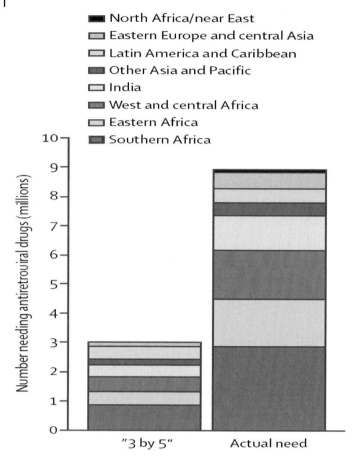

Figure 8.3 Comparison of WHO's "3 by 5" target (set for 2005) and the actual global need for antiretroviral therapy. (Source: Anema et al., Lancet 364, 1034–1035, 2004).

in turn infers that one in three women who attend antenatal clinic would be infected. That represents a total of 1.8 million people – a simply phenomenal number. It is inconceivable how drugs can possibly be obtained on such a scale; moreover, the physicians are unfamiliar with the situation and need to be trained. Africa has never had these drugs, and so a supply line must be developed, although if that were to be broken – perhaps if resistance to a drug were to be developed – the situation would rapidly become catastrophic. A short-term break in supply for most health materials is merely problematic, but if antiretroviral supplies were to cease for two months then the problem would be virtually insurmountable.

The possible answer to this problem is that systems must be devised and medical and laboratory staff trained. Procedures which are routine in Europe

Table 8.4 Estimated antiretroviral therapy coverage and overall therapy needs in developing countries. (Source: WHO).

WHO region	Estimated number of people on antiretroviral therapy, June 2004	Estimated antiretroviral therapy need, 2004–2005	Antiretroviral therapy coverage
African Region	150 000	3 840 000	4%
Region of the Americas	220 000	410 000	54%
European Region	11 000	120 000	9%
Eastern Mediterranean Region	4 000	100 000	4%
South-East Asia Region	40 000	860 000	5%
Western Pacific Region	15 000	170 000	9%
Total	440 000	5 500 000	8%

and the United States, such as CD4 cell counts and viral loads are simply unaffordable in Africa. Nonetheless, there is a multitude of issues to put in place for antiretroviral supply, and South Africa has the infrastructure to achieve this. The medical personnel are present and could probably cope with the situation, but the main problem is one of scale.

Whilst it can be assumed that the laboratories are prepared correctly and that the potential problems of resistance can be overcome, a recent report has suggested that in Africa, over the next ten years, even the provision of antiretrovirals will not necessarily reduce the horizontal transmission of AIDS. Such doubt is based on three pieces of evidence, all of which are as yet unproven. There is an expectation that the incidence of AIDS will suddenly fall, although this is uncertain and most likely will not be noticed for the next ten years. It is also suggested that, during a course of treatment there is a 5% incidence of acquired resistance, and that transmission resistance (resistant viruses being transmitted from one person to the next) will be a problem. There is, therefore a clear need to set up public awareness systems, rather than to investigate the transmission of HIV that is resistant to antiretrovirals, and to improve the systems of care such that patients can be managed more scientifically. There is, in fact, an entire list of problems dealing with the implementation of antiretrovirals, the best approach being to develop a worldwide collaboration that will include adequate resources to provide antiretrovirals, yet not compromise preventive activities, which must clearly be continued.

Preventive activities, such as condom use and changes in sexual behavior, are not especially easy to develop, but are generally inexpensive to implement.

HIV Prevention Strategies

➤ High-level political leadership
➤ Multi-sectoral approach: civil/religious society
➤ Population-based programs to change social norms
➤ Open communication of sexual activities and HIV/AIDS
➤ Programs to combat stigma/discrimination
➤ Condom promotion
➤ Surveillance and control of sexually transmitted infections
➤ Interventions targeting key "bridge" populations
➤ Prevention of mother-to-child transmission
➤ Safe blood transfusion, harm reduction

Figure 8.4 HIV prevention strategies.

Today, it is unclear which of these procedures is the most important, and the situation is far from being resolved (Figure 8.4). Many different groups, including religious sects, government ministers, scientists and various communities, have different points of view as to why some success has been achieved. There is clearly a conglomeration of these opinions, but there is no single reductionist approach to solving the AIDS problem.

8.5
Is Biotechnology Succeeding in Africa?

The question remains as to whether biotechnology has served Africa's needs. In the past, this has been thought not to be the case, and the problem with the AIDS epidemic was considered as being largely due to the failure of access to antiretrovirals, with both governments and international agencies as the responsible parties. There is, however, a new effort being undertaken to provide the support needed, notably in the case of the AIDS epidemic which is simply overwhelming parts of Africa. Another – sadly, correct – accusation is that AIDS is diverting funds from all other health aspects, such as breastfeeding and vitamin A intake, for which almost no funds are available because it is so easy to obtain money to investigate HIV. The problem is then that a balance must be found for future investigations, although the future of Africa is clearly dependent upon technology to which Africa's people require increased access. They must be supported in making choices about what they do or do not want, and they need help in order to sustain whatever funding they obtain for the growth and development of those communities. There is particular concern about external funding, which may not be sustainable. However, it is likely that this scale of support is unprecedented, and the opportunity must not be missed.

Even the greatest technology in the world will not be able to solve certain major problems, and indeed Africa has major problems that impact upon the country and its peoples, and also on much of the developing world. Whilst popular concern suggests that modern global power relationships do not affect Africa, the reverse is in fact true. These relationships have fundamental effects, and Africa is itself demanding equal representation on the Security Council of the United Nations. Debate must also be ongoing with regard to the World Trade Organization and intellectual property. Since the 13th AIDS Conference, which was held in Durban, there has been an enormous – and successful – struggle to obtain access to antiretroviral drugs, mainly by pressurizing pharmaceutical companies, businesses and governments. In recent years, the cost of antiretrovirals has plummeted, with typical annual costs in the US of 20 000 US$ falling to only about 300 US$ in South Africa. This is indeed an affordable proposition. In the case of nevirapine, the pharmaceutical company has virtually provided it free of charge to poor countries. However, the cost of preventing mother-to-child transmission has moved beyond the area of drug access such that is now a question of costs for staff, infrastructure and antenatal care.

There could be no simpler and cheaper means of reducing mother-to-child transmission of AIDS than single-dose nevirapine. It is easier to administer than polio vaccine (oral tablets for the mother; oral doses for the baby), yet globally less than 20% of those women who require nevirapine actually have access to the drug. In South Africa – which is far wealthier – this figure may reach 50%. Such a situation can only be described as scandalous.

8.6
Conclusions

Today, in Africa, the battle to obtain cheaper drugs and to retain intellectual property rights forms part of daily life. The most awkward aspect of the battle is that it is being fought not against enemies but against friends, through the courts, by using democratic processes to provide antiretroviral drugs and better health services in general. These debates are important, but then so too is the culpability of African countries, the leaders of which, whilst disagreeing violently about many scientific issues, have done much to overcome the problems of conflicts, wars and forced and internal migrations. Technology cannot function in the presence of an unstable society, and good government is essential in order to promote the democracy and freedom that allows not only those in power but also the international community to be questioned on behalf of the people.

Author Biography

Peter Paradiso

Vice President, New Business Development, Wyeth

Peter R. Paradiso, Ph.D., is Vice President, New Business and Scientific Affairs for Wyeth Vaccines, a Division of Wyeth Pharmaceuticals in Collegeville, PA. In this position, he is responsible for global scientific affairs and strategic planning within the vaccine research and development group and for commercial oversight of products in development. He has worked in the field of vaccine development at Wyeth for the past 20 years. P. R. Paradiso served as a member of the National Vaccine Advisory Committee (NVAC) and is currently a member of the Advisory Council on Immunization for New York State and a liaison member of the CDC's Advisory Committee on Immunization Practices (ACIP). He has published broadly in the field of pediatric vaccines, especially in the areas of glycoconjugates, combination vaccines and respiratory viral vaccines. P. R. Paradiso has been involved in the development and the global registration of vaccines for H. influenzae type b, rotavirus, Neisseria meningitis group C, Streptococcus pneumoniae and influenza. He has also served as an advisor to the WHO's Strategic Advisory Group of Experts on vaccines and to the GAVI Task Force on Research and Development.

9
New Vaccines with Global Impact: The Issue of Access

Peter Paradiso

9.1
Introduction

In this chapter, attention will be focused on the development of new vaccines rather than improving the access to available vaccines, as has been described in some of the previous chapters in this volume.

There are many issues involved, the majority of which are ethical in nature. In general, the vaccine market is one of most difficult to penetrate because experience is limited, and this presents some of the greatest challenges. Some years ago, in a movie called *Field of Dreams*, the main character built a baseball field, following the slogan of "… build it and they will come". The baseball field was built, and the people did indeed come. Today, this same philosophy is used for some vaccine developments, and this surely is indicative of the confidence expressed in vaccines, and in their market place.

The vaccine market place is an environment of great uncertainty, however – using the analogy with *Field of Dreams* – the question is whether they will come and, if they do, how many, when, and how to build and to prepare for that. In Chapter 7, Philippe Kourilsky referred to the "90–10 gap" as it applied to existing vaccines, as well as to vaccines under development. It is disconcerting to see such a high disease burden for measles when a vaccine has been available for 30 years but has simply not been taken up. When considering what production capacity should be built or considered for new vaccines, and how they should be tested, it is perhaps difficult to see the route ahead because of the unsure delivery mechanisms involved.

Health for All?: Analyses and Recommendations
Edited by The World Life Sciences Forum – BioVision
Copyright © 2005 Wiley-VCH Verlag GmbH & Co. KGaA, Weinheim
ISBN: 3-527-31489-X

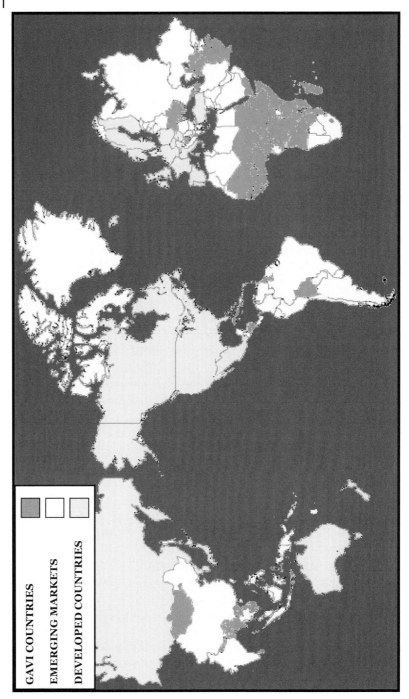

Figure 9.1 Countries covered by the Global Alliance for Vaccines and Immunization (GAVI).

9.2
Vaccine Development

There are two main areas involved in vaccine development. It is clear that issues of impact have been important, whilst establishing value is also important in the development of vaccines and testing of their efficacy. Building awareness, country by country, region by region, is critical in all market places, and again participation is via collaboration, as is the case with epidemiology and vaccine studies. To determine demand and to ensure supply is very difficult within this environment, but these must be undertaken in order to help predict an assurance of delivery systems and infrastructures.

The map in Figure 9.1 highlights the market places for vaccines worldwide. Of the three market places considered, those of the developed countries are where attention has traditionally been focused. Then, there are the emerging markets – these are quite large and have economies somewhere between those of the developing country markets and the United States and Europe. Finally, there are the least developing or poorest countries, with per capita incomes of less than US$ 1000 per year. The problem here is how to balance demand and production between these market places.

From the perspective of size, the developed and the least-developed markets are quite similar, whilst the less-developed markets are slightly larger, though not by orders of magnitude. So, it sometimes becomes less of an issue of whether the capacity can be built up, but more of understanding how much capacity is needed and what the demand will be.

9.3
Vaccine Supply

There must first of all be a partnership in determining the answers to some of these issues for all of the vaccines under development, and that partnership must include vaccine development and supply. Generally, the manufacturers, the donors and the financiers of those vaccines and the countries must be involved in the access and the delivery. Facilitating that are groups such as the Global Alliance for Vaccines and Immunization (GAVI). Recently, these have become highly effective, and have formed important working teams to focus on specific vaccine areas such as rotavirus, meningococcal disease and pneumococcal disease. The aim of this is to advance vaccine access, because past experience shows that it takes a long time for vaccines to be transferred from developed markets to developing markets. Figure 9.2 shows the situation for two vaccines in low-income countries, from the time of launch onto the developed market. *Haemophilus influenzae* B vaccine was introduced in the late 1980s/early 1990s in the developed world, in the United States and in

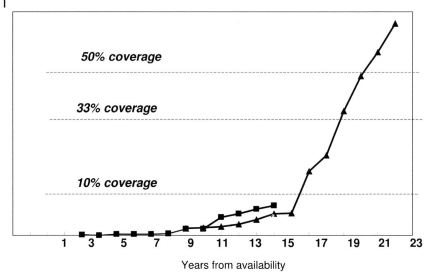

Figure 9.2 Delay in the use of vaccines in low-income countries.
Coverage data are for the 75 lowest income countries.
▲ hepatitis B vaccine; ■ H. influenzae B vaccine.

Europe. Today, 15 years later, the vaccine still has not been successfully introduced everywhere, despite its value having been recognized. The delivery simply has not happened. The GAVI has stated that their goal is to enhance uptake, so that better access to vaccines is achieved more quickly. However, from the viewpoint of the vaccine manufacturer, it is difficult to identify in which direction the route should progress – where to manufacture the vaccine, at what level, and at what point in time.

9.4
Disease Burden and Vaccine Efficacy

The relationship between disease burden and vaccine efficacy is referred to as "establishing value", and includes the issue of clinical trials. It also involves where and when such trials are conducted, the ethics involved, and the impact that these trials will have. The accusation is often leveled that clinical trials are not conducted in developing countries simultaneously to their being conducted in developed countries; that our focus is centered on the developed countries because that is where the market places are. However, this is not always true, and trials have been conducted and efforts made to generate data that will be available in a variety of market places. In either case, it does not appear to influence the issue of access.

In recent years, there are three notable vaccines that have been prepared, these being directed respectively against *H. influenzae* B, rotavirus and *Streptococcus pneumoniae* (Table 9.1). *H. influenzae* B is reported to be responsible for 400 000 deaths each year. Rotavirus and *S. pneumoniae* cause serious diseases, but in general, the primary efficacy trials are performed in developed country settings, and in the US and Europe in particular. Subsequently, their value and the implications of clinical trials performed in developing countries, either pre-licensure or post-licensure, becomes an issue of vaccine development. Moreover, there are ethical considerations as to whether this occurs pre-licensure or post-licensure, in addition to certain responsibilities for access after the trials have been completed.

Table 9.1 New vaccines with global impact.

Pathogen	Global health burden [deaths per year]	Vaccine development history
H. influenzae B	400 000	Pre-licensure efficacy trials in US and Finland
		Licensed for infants in the US and EU (1990)
		Post-licensure efficacy trial in Chile (1992–1995)
		Post-licensure efficacy trial in The Gambia (1993–1995)
Rotavirus	440 000	Pre-licensure efficacy trials in the US and Finland
		Pre-licensure efficacy trials in Venezuela
		Efficacy trials of prototype vaccine in Peru and Brazil
		Licensed in the US (1998) and EU (1999), but withdrawn from market in 1999
		Little interest post-withdrawal in developing countries
S. pneumoniae	> 1 000 000	Pre-licensure efficacy trial against IPD in the US
		Licensed in the US (2000) and EU (2001)
		Post-licensure efficacy trial against IPD and pneumonia in South Africa and The Gambia

IPD = invasive pneumococcal disease.

9.4.1
Haemophilus influenzae B

The *H. influenzae* B vaccine was licensed in the US and in Europe in infants in 1990, and efficacy trials were conducted predominantly in the US and in Finland. The efficacy was found to be very high for these vaccines, and approached 100% in many cases. In vaccinology terms, *H. influenzae* B has a saccharide on its outer shell against which an antibody can be induced. When this occurs, the bacterium is killed and the vaccine will be effective. The clinical trial showed this suggestion to be true. The next step was that if an antibody to the saccharide could be induced in a population, then the vaccine could be confidently expected to be effective. However, the issue is then whether it is ethical to continue with placebo-controlled trials when the vaccine's efficacy is virtually certain. There are both pros and cons to this approach, and in many cases it is questionable. It came as no surprise when, in subsequent randomized trials in Chile, the *Haemophilus* conjugate vaccine demonstrated the same level of efficacy. A subsequent trial in The Gambia showed vaccine efficacy against invasive disease. This was an example of a trial being conducted in a developing country and at the same time adding to the knowledge base, because surprisingly there was seen to be a significant reduction in all-cause pneumonia from *H. influenzae* B – a benefit which had not previously been shown.

If conducting a clinical trial in a developing country can add to the current information base, then the trial's outcome can be deemed valuable. However, the outcome with *H. influenzae* B had not been anticipated, so the prediction of a vaccine's value may be difficult. When, in general, the tenet is that "more information is better", the ethics of conducting trials in developing countries become slightly blurred when important new findings sometimes result from these.

9.4.2
Rotavirus

With RotaShield, the quadrivalent rotavirus vaccine that was licensed in the US during the late 1990s, pivotal trials were performed again in the US and Finland. There is in fact a pattern here, that Finland and the US seem to be excellent locations for efficacy trials! Equivalent vaccine efficacy pre-licensure trials were performed in Caracas, Venezuela within a developing country setting, whilst efficacy trials of prototype vaccines prior to RotaShield release were conducted in Peru and Brazil. Many of the trial outcomes were similar, with RotaShield showing very significant efficacy against severe disease. Thus, at the time of licensure in the US and in Europe, there was a robust efficacy profile for this vaccine in a variety of environments in both developed and developing country settings. The experience at the time of licensure, when communicating with health communities in Asia and Africa, was that further trials were required and that these should be conducted in the communities' regions. In Asia and Africa there was dissatisfaction at the level of efficacy, despite the manufacturers having embarked on Phase I (safety and immunogenicity) and Phase III (efficacy) trials before RotaShield could be introduced into developing countries. Thus, despite having conducted pre-licensure trials in a setting that was comparable to that in many developing countries, the vaccine was deemed unacceptable for these countries. In the US, there was also an issue with intussusception, an adverse event that occurred in about 1 in 10 000 patients, and consequently the recommendations for vaccine use were withdrawn. The clinical trials were first put on hold and then halted by the regulatory agencies. The message was clear – if there is a safety issue that leads the US and Europe to reject use of the vaccine, it will not be acceptable for developing countries, even if there is a fundamentally different risk/benefit ratio in these countries. Following the safety profile problems of intussusception, there was minimal interest in this vaccine and so it was not marketed, regardless of its potential benefit.

9.4.3
Streptococcus pneumoniae

The most recent experience was with *Streptococcus pneumoniae*. A seven-valent conjugate vaccine (Prevnar) was licensed in the US in 2000 and in Europe in 2001, and showed very high efficacy against invasive pneumococcal disease. As with the *Haemophilus* vaccine and meningococcal conjugate vaccines, this efficacy was correlated with an ability to induce an antibody response to the saccharides of the conjugate. During the course of the development, efficacy trials were initiated in several developing countries, including South Africa and The Gambia. These trials spanned the licensure period and, on completion, demonstrated efficacy not only against invasive disease but also – importantly – against pneumonia. Again, this is an example of efficacy trials that had a different endpoint which was focused on the needs of the population, in order to establish value among the population where the trial was conducted. In the US, the primary concern was invasive disease in meningitis, whereas in Africa the major concern was pneumonia. These were the endpoints of the trial. The vaccine was a nine-valent conjugate; it had the seven types that were in Prevnar, plus types 1 and 5 that are prevalent in Africa. The most important point of the trial was to determine whether, when serotypes were added, the efficacy remained. The trial would also confirm whether the vaccine was effective in the setting of urban Africa and under the EPI schedule. The endpoints were invasive disease, all X-ray proven pneumonia, admissions to hospital for any cause, and death from any cause. The results were quite

Table 9.2 The Gambia pneumococcal vaccine trial.
The vaccine used in this trial was a 9-valent conjugate vaccine.

Endpoint	Efficacy [% reduction]
Invasive disease	77[a]
X-ray-positive pneumonia	37[a]
Hospital admission	15[a]
Death	16[a]

[a] Statistically significant reduction.

outstanding, with a 77% reduction in invasive disease (which perhaps was expected based on other trial outcomes), as well as a reduction in all X-ray confirmed pneumonia of 37%, hospital admissions of 15%, and mortality of 16% (Table 9.2).

9.5
Ethical Perspectives

Two important issues arose from this trial, from an ethical perspective. The first issue was that although mortality was taken as a primary endpoint, during the course of the trial it became clear that many trial operatives were uncomfortable about counting and measuring deaths. Consequently, mortality was used as a secondary endpoint. However, despite doubts related to the power of the trial, it transpired that so many people died as a result of pneumococcal infection, mortality could indeed be used as an endpoint.

The second issue, which was addressed during the middle of the trial, was whether it was ethical to continue with a vaccine that had already been licensed in the US and Europe and was clearly effective. The decision was made to continue the trial, on the basis of the pneumonia endpoints and that new data would be obtained. The question is therefore, if efficacy has been demonstrated in one population, is it ethical to perform further placebo-controlled trials? In many cases the answer to this is yes, but only if fresh data are derived and new information is provided (Figure 9.3). The important point from an ethical perspective is what is required after the trial has been completed.

Two considerations spring to mind. The first concerns the placebo group if the trial is successful: in clinical trials involving efficacy, vaccine is often subsequently offered to the placebo population, and indeed this is often required in the protocol. This is no different from studies conducted within developing countries, although on occasion the method of vaccine delivery presents a

Is it ethical to do further placebo-controlled trials if efficacy of a vaccine has already been demonstrated in one population?

Pre-trial considerations
➤ Population effects on efficacy may be discovered
➤ New, perhaps more relevant, endpoints may be studied
➤ How likely is the population to have access to the vaccine if no trial is conducted?
➤ Will a trial have a positive impact on the subsequent introduction of the vaccine?

Post-trial requirements
➤ Vaccinate the placebo population
➤ Ensure access of the general population to the vaccine

Figure 9.3 Ethical issues associated with clinical trials in developing countries.

major challenge. On completion of the trial, there is also an obligation with regard to providing access to the trial vaccine, or to a vaccine that has already been tested. In a recent report that was posted on the Science and Development Network website (www.scidev.net, October 2003) B. Greenwood and W. P. Hausdorff spoke about post-trial access and what should be required. Their suggestion was that the manufacturers offer to supply the vaccine at a discounted, "affordable price" which reflects the economy of the country or region. The Minister of Health would then decide whether, at an affordable price, the country was interested in utilizing the vaccine and would be willing to take on the responsibility for the community's health. The third option would be that the Minister of Health was not be willing to make such a contribution, and the venture would be deemed not sufficiently important to proceed.

These points are easy to note down, but are difficult to deal with. The idea of the manufacturer donating vaccines to populations in which clinical trials have been conducted has been considered as a "cop-out", as it removes the potentially problematic issue of deciding an affordable price, and of deducing a long-term price for developing countries in general. Whether this approach might provide a sustainable mechanism for the country receiving free vaccine is unclear, but – somewhat surprisingly to me – as a company we have consistently been discouraged to do so during these studies.

9.6
Vaccine Demand and Capacity

If clinical trials are conducted successfully within a developing country setting, the next major decisions to be taken are related to adequate vaccine supply and confirmation of an ability to meet the demand generated. There are three market segments to examine, the first of which is to establish an affordable price. Discussions on price take the form of a circular discussion: How much vaccine is required? What is the cost? How can the cost be deduced until the demand is known? When is the vaccine required? In order to minimize this situation, the GAVI has set up mechanisms to ensure much more direct interaction to predict and understand both demand and timing, and to relate these properties to cost. The subject of cost and building up of capacity can be considered in two ways. The first way is to use existing capacity and to better utilize existing facilities, whilst the second way is to build new capacity. The latter approach represents a much greater challenge because it is associated with the prediction of demand. The difficulty here is that it takes five years to build and license an increased capacity. So, it is vital that the future demand is known, as it would be disastrous for new capacity to be idle due to a lack of demand.

The concept of alternative formulations and presentations for different market places must also be considered. Whereas in developing countries the market is predominantly multidose, in developed countries vaccine administration is invariably single-dose vials, without preservatives. Presentation – which seems so trivial – becomes an entire manufacturing regulatory process that must be traversed in order to develop different formulations for developing countries. In this respect, the manufacture of vaccines with preservatives becomes another issue. It is, therefore, critical to have third-party collaboration, and the partnerships created in developing countries, linked with the capacity build-up there, will probably be vital for the success of the vaccine manufacturers.

9.7
Affordability and Sustainability

It is important to recognize and to determine what is affordable and what is sustainable. It is likely that tiered pricing represents the only route to take, with the least-developed countries paying by far the least amount. This situation must be viewed by companies as being philanthropic – there is no other way to view it. Whether the vaccine is donated or a charge of a few dollars is made is irrelevant; the cost will inevitably be far less than that in a different setting, and may often be many-fold less than the production cost. The fact that such a gesture can be made, and without too much pain, is probably a good enough reason to do it. However, it must be understood by all concerned that to charge less in some markets does not mean that the same price is applicable worldwide.

Access to vaccines requires close collaboration between partners, and it must be emphasized that during the past five years the environment created by the GAVI, with all their effort, has become a very pleasant workplace. Ethical issues must be addressed by all partners, and the risks to a company's shareholders must be both identified and expressed. Moreover, the philanthropic nature of supply must be acknowledged and the need to tier pricing to match markets must be accepted.

General Bibliography and Suggested Reading

1 Africa Harvest Biotech Foundation International (AHBFI) website, F. Wambugu, *Scientist dares African media, Take up Gates' $200m opportunity.* http://www.ahbfi.org/newspaper/thirdquart2.htm.

2 BIO (Biotechnology Industry Organization) website, D. Eramian, *Patents save lives.* http://www.bio.org/speeches/speeches/20040624.asp.

3 Campaign for access to essential medicines website, *Médecins sans frontières.* http://www.accessmed-msf.org.

4 P. Farmer, *Infections and inequalities: the modern plagues*, Updated edition with a preface (**2001**), University of California Press.

5 P. Farmer, *Pathologies of power: health, human rights, and the new war on the poor* (**2003**), University of California Press.

6 L. R. Kass, *Life, liberty and the defense of dignity: the challenge for bioethics* (**2002**), Encounter Books, San Fransisco, California.

7 Nuffield Council on bioethics website, *The ethics of research related to healthcare in developing countries* (**2002**). http://www.nuffieldbioethics.org/go/screen/ourwork/developingcountries/publication_309.html.

8 Conference on Ethical Aspects of Research in Developing Countries, Ethics enhanced: Fair Benefits for Research in Developing Countries (**2002**), *Science* 298, 2133–2134.

9 A. R. Fooks, Development of oral vaccines for human use (**2000**), *Curr. Opin. Mol. Ther.* 2, 80–86.

10 G. Vogel, Malaria: A Complex New Vaccine Shows Promise (**2004**), *Science* 306, 587–589.

11 World Bank Institute website, J.-E. Aubert, *Promoting innovation in developing countries: a conceptual framework.* http://www-wds.worldbank.org/servlet/WDSContentServer/WDSP/IB/2005/02/23/000090341_20050223150015/Rendered/PDF/315100HNP0Phar1ng0Analysis01public1.pdf.

12 World Health Organization website, WHO and UNAIDS, *Guidance on ethics and equitable access to HIV treatment and care.* http://www.who.int/ethics/resource_allocation/en.

Health for All?: Analyses and Recommendations
Edited by The World Life Sciences Forum – BioVision
Copyright © 2005 Wiley-VCH Verlag GmbH & Co. KGaA, Weinheim
ISBN: 3-527-31489-X

13 P. FARMER, N. G. CAMPOS, Rethinking medical ethics: a view from below (**2004**), *Developing World Bioeth.* 4 (1), 17–41.

14 K. S. KHAN, Epidemiology and ethics: the perspective of the Third World (**1994**), *J. Pub. Health Policy* 15 (2), 218–225.

15 R. BAYER, As the second decade of AIDS begins: an international perspective on the ethics of the epidemic (**1992**), *AIDS* 6(6), 527–532.

16 N. M. P. KING, G. HENDERSON, Treatments of last resort: informed consent and the diffusion of new technology (**1991**), *Mercer Law Rev.* 42(3), 1007–1050.

17 KUMAR S. HOPE, News, but little cheer, for India's revised code of ethics (**1998**), *Lancet* 351, 347.

18 L. LASAGNA, The Helsinki Declaration: timeless guide or irrelevant anachronism? (**1995**), *J. Clin. Psychopharmacol.* 15 (2), 96–98.

19 Center for Ethics and Humanities in the Life Sciences website, P. FARMER, Lecture: Ethics and Equity: Current Challenges in International Health (**2003**), *Medical Humanities Report* 25 (1). http://bioethics.msu.edu/mhr/03f/farmer.html.

20 M. GOOZNER, *The $800 Million Pill – The Truth behind the Cost of New Drugs*, The University of California Press.

21 P. FARMER, *AIDS and Accusation: Haiti and the Geography of Blame*, Comparative Studies of Health Systems and Medical Care, The University of California Press.

22 A. MULLINGS, Genetic research in the Third World (developing) countries – science or exploitation? (**2001**), *St Thomas Law Rev.* 13 (4), 955–964.

23 P. A. CLARK, AIDS research in developing countries: do the ends justify the means? (**2002**), *Med. Sci. Monit.* 8 (9), ED5-16.

24 M. H. KOTTOW, Who is my brother's keeper? (**2002**), *J. Med Ethics* 28 (1), 24–27.

25 World Medical Association website, Guideline: Proposed revision of the Declaration of Helsinki (**1999**), *Bull. Med. Ethics* 150, 18–22. http://www.wma.net/e/ethicsunit/pdf/draft_historical_contemporary_perspectives.pdf.

26 R. MACKLIN, International Research: Ethical Imperialism or Ethical Pluralism? (**1999**), *Accountability in Research: Policies & Quality Assurance* 7, Issue 1, 59–84. http://search.epnet.com/login.aspx?direct=true&db=buh&an=3984128.

27 L. ROBERTS, Vaccines: Rotavirus Vaccines' Second Chance (**2004**), *Science* 305, 1890–1893.

28 H. TRISTAN ENGELHARDT, *The foundations of bioethics*, Oxford University Presse (**1996**).

Key Messages

Health for All?: Analyses and Recommendations
Edited by The World Life Sciences Forum – BioVision
Copyright © 2005 Wiley-VCH Verlag GmbH & Co. KGaA, Weinheim
ISBN: 3-527-31489-X

Author Biographies

Gerald T. Keusch

Assistant Provost, Medical Campus; Associate Dean, School of Public Health, Boston University; former Director, Fogarty International Center, NIH

Gerald T. Keusch is Assistant Provost for Global Health, Boston University Medical Campus and Associate Dean for Global Health at Boston University School of Public Health. Prior to this appointment, he served as Director of the Fogarty International Center at the National Institutes of Health and Associate Director for International Research in the office of the NIH Director. A graduate of Columbia College and Harvard Medical School, he is Board Certified in Internal Medicine and Infectious Diseases. He has been involved in clinical medicine, teaching and research for his entire career, most recently as Professor of Medicine at Tufts University School of Medicine and Senior Attending Physician and Chief of the Division of Geographic Medicine and Infectious Diseases, at the New England Medical Center in Boston, MA. His research has ranged from the molecular pathogenesis of tropical infectious diseases to field research in nutrition, immunology, host susceptibility, and the treatment of tropical infectious diseases and HIV/AIDS. He was a Faculty Associate at Harvard Institute for International Development in the Health Office. Dr. Keusch is the author of over 300 original publications, reviews and book chapters, and he is the editor of 8 scientific books. He is the recipient of the Squibb, Finland and Bristol awards for research excellence of the Infectious Diseases Society of America, and has delivered numerous named lectures on topics of science and global health at leading institutions around the world. He is presently involved in international health research and policy with the NIH, the U.S. National Academy of Sciences' Institute of Medicine, the United Nations, and the World Health Organization. Under his leadership, the programs of the Fogarty International Center were greatly expanded and focused on the creation of a global culture of science and to harness science for global health. Fogarty now supports research, capacity building, and science policy on the pressing global issues in infectious diseases, the growing burden of non-communicable diseases, and critical cross-cutting social science issues such as the ethical conduct of research, intellectual property rights and global public goods, stigma, the impact of improved health on economic development and the effect of economic development on the environment and health.

Kul Chandra Gautam

Deputy Executive Director, UNICEF

Kul Chandra Gautam is currently Assistant Secretary-General of the United Nations and Deputy Executive Director of the United Nations Children's Fund (UNICEF) at its Headquarters in New York. He is responsible for providing leadership in strategic planning, programme development, resource mobilization, and promoting partnership for children and development among UN agencies, donors and civil society organizations.

Mr. Gautam has had a long and distinguished career with UNICEF. Starting in 1973, he served as Programme Officer in Cambodia and Indonesia, as UNICEF Country Representative in Laos, Haiti, and India, and as Regional Director for Asia and Pacific. He also served as Chief for Latin America and the Caribbean, as Director for Planning, and Director of Programme Division at UNICEF Headquarters in New York.

As Director of Programme Division and Acting Deputy Executive Director (Programme), Mr. Gautam had major responsibility for developing and overseeing policy and programme strategies for UNICEF cooperation in developing countries in the early 1990s.

He was the key senior UNICEF officer responsible for drafting the Declaration and Plan of Action of the 1990 World Summit for Children, the largest gathering of world leaders in history until that time. In May 2002 he led the organization of another major United Nations conference, the Special Session of the General Assembly on Children attended by 70 world leaders and thousands of child rights activists and civil society leaders, including celebrities and Nobel Prize Laureates.

Mr. Gautam is a citizen of Nepal. He received his higher education in the United States of America, at Dartmouth College, Princeton University, and Harvard University. Mr. Gautam is married with a daughter and a son.

George R. Siber

Executive Vice President and CSO, Wyeth Vaccines Research

George R. Siber joined Wyeth Lederle Vaccines as Vice President and Chief Scientific Officer in August, 1996. He became Senior Vice President in August, 1999 and Executive Vice President in June, 2002. In this capacity he is responsible for discovery research in bacterial vaccines, viral vaccines, immunology and genetic vaccines, process and analytical development, clinical development, and scientific affairs for Wyeth Vaccines Research.

While at Wyeth Dr. Siber has overseen the development and approval of an acellular pertussis vaccine for infants (Acel-Imune), a vaccine to prevent Rotavirus diarrhea in infants (RotaShield), a glycoconjugate vaccine to prevent group C meningococcal meningitis (Meningitec), a 7 component glyco-conjugate vaccine to prevent pneumococcal disease in infants and children (Prevnar), and a cold adapted nasally administered influenza vaccine in collaboration with MedImmune (FluMist).

Prior to joining Wyeth Dr. Siber was Director of the Massachusetts Public Health Biologic Laboratories and Associate Professor of Medicine with the Harvard Medical School, Dana Farber Cancer Institute. During this time he oversaw research on acellular pertussis and Hemophilus influenza vaccines, the development and approval of CMV Immune Globulin (Cytogam®) and RSV Immune Globulin (Respigam®) and the production of DTP vaccines and immune globulins for the State of Massachusetts.

Dr. Siber's research interests have included the evaluation of the human immune response to polysaccharide and protein antigens, the development of vaccines and immune globulins against Hib, pneumococci, meningococci, pertussis and RSV and maternal immunization to prevent perinatal and early neonatal infections. He has authored more than 150 scientific articles in peer-reviewed journals. He holds 3 issued patents which support a licensed diagnostic test for meningitis (Bactigen®) and an antibody based preventative for respiratory syncytial virus infections in high-risk children (Respigam®).

Dr. Siber has served on numerous advisory committees including the WHO/ UNDP Steering Committee for Encapsulated Bacterial Vaccines, the Steering Committee for Development of Streptococcus Pneumonia Vaccine for the Pan American Health Organization, the Institute of Medicine Committee on the Children's Vaccine Initiative, the NIH Blue Ribbon Panel for Bioterrorism and its Implications for Biomedical Research, Chairman of the review of the US Army's HIV research program, and the Board of Scientific Counselors for the National Vaccine Center.

Synthesis and Recommendations

Gerald T. Keusch, Kul Chandra Gautam, and George R. Siber

1
Basic Facts

The global situation with regard to vaccines and vaccination can be characterized as follows:

- 75% of children are immunized, but 36 million newborns remain without access to basic vaccines.
- Between 2 and 3 million children will die from a vaccine preventable disease each year. This approximates to 250,000 deaths per month = 1 "Invisible Tsunami" per month.
- In the last 5 years, global action such as the GAVI initiative has reached 45 million more children with new vaccines and prevented 700,000 deaths.
- The Biotechnology revolution opens new avenues for vaccine discovery.
- Vaccines have taken 15–20 years to reach developing countries after their first introduction in developed countries.

2
Key Concerns

Key concerns in the global health sector for vaccine development and vaccination have been identified that must be overcome to improve the current situation.

- The true value of vaccines is not appreciated.
- Infrastructure deficiencies continue to impede access in developing countries.
- Funding is not sustained over time; one reason for this is donor fatigue.

Health for All?: Analyses and Recommendations
Edited by The World Life Sciences Forum – BioVision
Copyright © 2005 Wiley-VCH Verlag GmbH & Co. KGaA, Weinheim
ISBN: 3-527-31489-X

- There is a lack of political commitment because benefits from vaccination are delayed as compared to other health initiatives.
- Public trust in vaccination is lost.
- Technical challenges remain daunting for high priority vaccines like HIV, tuberculosis and malaria.
- Increasing aversion to risk has led to increasing regulatory barriers and increasing costs of development.

3
Call for Action

Action points have been identified to address the key concerns as given above. Actions needed are different with regard to vaccines already in use (old vaccines), those that are ready to be used (recent vaccines), and those that are at present not available (vaccines to be developed).

3.1
Old Vaccines

These are vaccines that are fully developed, have an established record of safety and efficacy, where supply is abundant, and the price is not the issue. Among this group are vaccines against, e.g., measles, pertussis, tetanus.

- Access is the critical barrier.

- To improve access, multiple partnerships are needed.

- For a vaccination program to be successful, commitment must be maintained by each partner:
  - Local governments must provide infrastructure, staffing and ensure a working cold chain.
  - Global Initiatives like GAVI and the Gates Foundation must provide funding.
  - Vaccine manufacturers must provide adequate vaccine supply, including low-cost generics.

3.2
Recent Vaccines

These are vaccines that have been recently developed. They are available but not used. Among this group are vaccines against, e.g., rotavirus and pneumo-coccus.

- High price is required to recover development costs and capital expenditure on manufacturing facilities, unless one or more of the following criteria can be met:
  – Advanced purchasing commitments are signed.
  – There is help from an International Financing Facility.
  – Three-tiered pricing is introduced.

- Access needs to be scaled up

3.3
Vaccines to Be Developed

These are vaccines that have not yet been developed, but where a major need exists. Among this group are vaccines against the "big three", malaria, tuberculosis, and HIV.

- Scientific breakthroughs are needed (compare the Gates Grand Challenges in Global Health).
- Cost of vaccine development must be reduced, e.g. by lowering regulatory barriers.
- Public investment must be increased, e.g. by creating a Global Fund for Tropical Disease.
- Public awareness and acceptance of the required investment must be increased.